T0211377

The Electric Car

Oliver Schwedes · Marcus Keichel
Editors

The Electric Car

Mobility in Upheaval

Springer

Editors
Oliver Schwedes
Integrated Transport Planning
TU Berlin
Berlin, Germany

Marcus Keichel
Läufer+Keichel
Berlin, Germany

ISBN 978-3-658-29759-6   ISBN 978-3-658-29760-2   (eBook)
https://doi.org/10.1007/978-3-658-29760-2

This Springer imprint is published by the registered company Springer Fachmedien Wiesbaden GmbH, part of Springer Nature.
The registered company address is: Abraham-Lincoln-Str. 46, 65189 Wiesbaden, Germany

Contents

1 Introductory Remarks 1
Oliver Schwedes and Marcus Keichel

2 Dominion Over Space and Time 9
Wolfgang Ruppert

3 Object of Desire 41
Oliver Schwedes

4 Completely New Possibilities 67
Marcus Keichel

5 The Benchmark Is Still Current Behavior 95
Christine Ahrend and Jessica Stock

6 "Focus Battery" 115
Henning Wallentowitz

7 In Place of an Afterword 141
Claus Leggewie

Correction to: The Electric Car C1
Oliver Schwedes and Marcus Keichel

Bibliography 145

The original version of this book was revised The correction to this book is available at
https://doi.org/10.1007/978-3-658-29760-2_8

List of Contributors

Christine Ahrend is the First Vice President for Research, Appointment Strategy, Knowledge & Technology Transfer and deputy to the president at the Technical University of Berlin.

Marcus Keichel is an industrial designer and was a visiting professor at the Institute for Product and Process Design at the Berlin University of the Arts from 2007 to 2010.

Claus Leggewie has the Ludwig Börne-Professorship at the Justus Liebig-University of Göttingen heading the "Panel on Planetary Thinking".

Wolfgang Ruppert heads the Center for Cultural History Studies at the Berlin University of the Arts.

Oliver Schwedes is head of the Integrated Transport Planning Department at the Technical University of Berlin.

Jessica Stock was a research assistant at the Department of Integrated Transport Planning at the Technical University of Berlin and a doctoral fellow in the DFG Research Training Group "Innovation Society Today: The Reflective Production of the New" until 2016.

Henning Wallentowitz headed the Institute of Automotive Engineering Aachen (IKA) at RWTH Aachen University until 2008.

Introductory Remarks

1

Plea for a New Mobility Culture

Oliver Schwedes and Marcus Keichel

Introductory Remarks

The electric car is coming. The German government decided to support the expansion of electric transport beyond 2012 with millions of euros in funding—one million electric cars are meant to be on German roads by 2020.

A lot has happened since the first funds from the *Economic stimulus package II* were released, which got the *National Electromobility Development Plan* underway. The German automotive industry started developing electric cars and was promoting its concepts even before sales started. It promised that, by the end of 2014, 15 different German-produced models would be available. The topic is also prominent in the media. The national newspapers in particular regularly report on the activities of the participating actors from politics, research and industry. However, after initial euphoria, the reporting has become increasingly dispassio-

---

The original version of this chapter was revised. A correction to this chapter is available at https://doi.org/10.1007/978-3-658-29760-2_8

O. Schwedes (✉)
Integrated Transport Planning, TU Berlin, Berlin, Germany
e-mail: oliver.schwedes@tu-berlin.de

M. Keichel
Läufer+Keichel, Berlin, Germany
e-mail: marcus.keichel@laeuferkeichel.de

© Springer Fachmedien Wiesbaden GmbH, part of Springer Nature 2021, corrected publication 2021
O. Schwedes, M. Keichel (eds.), *The Electric Car*,
https://doi.org/10.1007/978-3-658-29760-2_1

1

nate and sometimes sceptical. At the beginning of 2013, for example, it was reported that from January to November 2012 only 2695 electric cars were registered in Germany—equivalent to 10% of the number planned by the industry. At about the same time it became public that the premium manufacturer Audi had discontinued its development projects in view of the sluggish demand for electric cars. The German Association of the Automotive Industry expected that a "boost in the market with higher unit sales" would not occur until the second half of the decade.[1] So the electric car will come, but at the moment it is open as to whether and how quickly it will prevail in larger numbers and thus become a serious carrier of our mobility culture. The question is: on what does this outcome depend?

Actors from Politics, Research and Industry

Picture 1.1 Elektricity Berlin [IVP]

---

[1] *Süddeutsche Zeitung*, 12–13 January 2013.

Electric Car as Carrier of Our Mobility Culture

Electric transport, i.e. the electric car including the associated infrastructure, is a complex matter, with various participants endeavoring to bring it about. The mere fact that participants from fields as diverse as research, politics and business are involved in shaping the process and contributing their perspectives and interests obviously entails the potential for friction. Even within the individual disciplines unqualified consensus is the exception to the rule, and by their very nature different approaches compete with each other. Against this backdrop, there is not only uncertainty regarding *whether* the electric car will establish itself as a vehicle of mobility culture but also concerning *how* it will do so. Quite apart from this state of uncertainty, it is becoming apparent that the current debate on electric transport is strongly positivistically oriented. In many cases there is a belief that the electric motor will sooner or later either supplement the combustion engine (hybrid technology) or simply replace it. Consumers would thus gradually switch from the conventional car to the electric car without major changes or even restrictions in the use of the car as such. Publications on the matter are characterized by a belief in technological progress, which can supposedly resolve the conflict between the need to conserve resources and unlimited individual mobility. It is obvious why this position is dominant: the promise that 'technology' will resolve the problem and that people in the highly developed societies can go on living as before is clearly extremely attractive (see Schwedes, Chap. 3).

Conflict of Goals Between Resource Conservation and Individual Mobility

This volume takes up this point. The authors have their doubts about a one-sided standpoint that places its trust in progress and instead assume that the initiative for electric transport can only be successful in the sense of meeting its ecological objectives if it is accompanied by a process of political and cultural reforms. In this perspective, a comprehensively revised energy policy (renewable energies) and a change in individual mobility behaviour appear to be necessary criteria for the initiative to be successful.

The transport sector in particular has impressively demonstrated that technological progress and economic prosperity do not automatically lead to sustainable development. Although technological innovation has been successful for decades, for example in the construction of ever more efficient engines, transport is the only sector today in which CO_2 emissions will continue to rise. Efficiency gains from technological innovations are repeatedly more than counteracted by the absolute growth in traffic volume, which continues to this day. In order to avoid a similarly contradictory development in electric transport, there is an obvious need for courage in the sphere of political regulation, even when this involves highly unpopular measures such as the reassessment of subsidies for private transport in favour of other modes of transport.

Primacy of Politics

Last but not least, historical developments make it clear that real progress can only be achieved if the primacy of politics over the economy is asserted and if the political will to shape society has an impact on long-established social traditions. Thus, for example, the introduction of the sewage system for private houses at the end of the nineteenth century—today considered an undisputed achievement of civilization—was part of the political clashes in the context of the hygiene movement and was accompanied by fierce conflicts. For various reasons, people resisted intervention in their everyday lives, which ultimately delayed the implementation of technical innovation for decades. At the time it was the city authorities who imposed the sewage system for the common good and against the massive resistance of the population.

Achievement of Civilization

However, there are also recent examples that illustrate the power of politics to act in the common good. Who would have thought that a Europe-wide smoking ban could be imposed overnight? A widely established cultural practice was thus quickly banished from the public sphere by politicians for the benefit of the general public—in this case, however, without having to overcome extreme resistance. In contrast to the historical example of sewage disposal, the population in this case, contrary to the prevailing impression, was obviously mentally prepared for the change. Here, politicians were assigned the task of finally bringing a decade-long process of enlightening civil society to a collectively binding decision. The politically-induced phase-out of nuclear power would also be worthy of consideration in this context. As contradictory as it is, this process—initiated by the Democratic Socialist-Green Party government in the late 1990s and ultimately confirmed by the Christian Democratic-Liberal government in the wake of the

Japanese nuclear catastrophe in 2011—impressively underscores the potential scope for political action. At the same time, the phase-out of nuclear power marks a shift in energy policy towards renewable energies, which are nothing short of an indispensable criterion for the ecological success of the electric car.

Political Control and Enlightenment

Since in a democratic society changes—for example in the mobility behaviour of the population—cannot simply be imposed or enforced through far-reaching prohibitions, measures of political control must be accompanied by educational initiatives. A re-examination of the sometimes mythical cult of the car would be an important goal in this respect. In turn, for this to happen it is indispensable to undertake a cultural-historical, critical examination of the emergence and significance of this cult as well as the historical fixation on the car as a leading product of modern traffic development and as an object of conspicuous consumption (cf. Ruppert, Chap. 2). Furthermore, the critical examination of current design developments in the field of electric cars is of great importance. The symbolism created by the designers has a considerable influence on the nature of the emotional relationship to the product. The design of electric cars will play a decisive role in determining whether it will be possible to temper the mental fixation on the car as an object of extravagant longings for power, speed and prestige and instead to create new and ultimately more humane meaning for the car (cf. Keichel, Chap. 4).

The Leading Product of Modern Traffic Development

We are convinced that the 'ecological question' today represents a challenge comparable to the 'social question' of the nineteenth century. As in the past, it is a question of value-related, comprehensive cultural change in the interest of the common good. The transport sector is of major importance for meeting environmental challenges. If it is to be successful, the introduction of electric transport must be linked to far-reaching cultural reforms of individual mobility behaviour (cf. Ahrend and Stock, Chap. 5). Due to the importance of the car as an economic object and the frequently demonstrated libidinous character of private car ownership, the actors involved must reckon with considerable resistance from interest groups as well as certain parts of the population, as was the case in the past and in other contexts. Ultimately, this resistance can only be attenuated through clearly articulated political resolve, by educating the public and through the development of positive alternatives.

The Ecological Challenge

The authors of this volume would like to make a contribution to this process. They argue that the decision to opt for electric transport should be used as an opportunity to critically examine all the experiences associated with the more than 100 year history of 'automobility'. The electric car, we argue, should be the starting point for a reform of the mobility culture of modern societies. The aim of this reform should be to alleviate the sometimes irrational cult of mobility and to encourage a better balance between the use of collective transport systems on the one hand and private cars on the other.

The Cult of Mobility

For this, the mobility form of 'being a passenger' in relation to 'being at the wheel oneself' would have to be emotionally re-evaluated. This may seem difficult, but here too historical developments make it clear that reinterpretations of this kind are possible in principle: for a long time horseback riding was associated with the strength and power of warriors and rulers. When in the sixteenth century "ceremonial carriages" were developed, the rulers changed from horses to carriages. Being driven around in a carriage was now regarded as a sign of a privileged position in society and was therefore more attractive than riding a horse. This perspective remained valid until the beginnings of auto-mobilisation: it is claimed that Gottlieb Daimler, for example, still assumed that a maximum of 5000 automobiles could be sold because at the time more chauffeurs were not to be found.

Tempering the Mental Fixation on the Automobile

If over a long period of time being a passenger was more highly esteemed than holding the reins oneself, it is conceivable that this could be the case again in the future. The decisive factor is the meaning that people assign to the respective form of mobility. As an expression of a modern lifestyle based on reason and responsibility, using public transport systems could become as important as conserving heating energy or separating waste. However, this would require an attractive range of collective transport systems and a tempering of the mental fixation of a large part of the population on the automobile. The car should not, however, be demonized or even abolished. Rather, it is a question of rediscovering its actual qualities by using it in a more moderate, less obsessive way.

There is some evidence that a less excessive and at the same time less frenetic use of the car increases rather than decreases the quality of life—quite apart from the positive ecological effects. If it were possible to combine the introduction of the electric car with a paradigm shift in mobility culture, the criteria by which the (everyday) suitability of electric cars would be assessed would shift. What has so far always been interpreted as a weakness in comparison with cars powered by an internal combustion engine may well prove to be a strength.

Dominion Over Space and Time

2

On the Cultural History of the Automobile

Wolfgang Ruppert

Introductory Remarks

2012 was considered to be the year with the highest gasoline prices in the history of the automobile. Nevertheless, the number of fuel-guzzling SUVs (sports utility vehicles) continued to rise to 16% of new car sales. Models such as the Porsche Cayenne enjoy great international popularity. This car combines the prestige of the Porsche sports car, established over many years, with the brawny form of a sedan, which rises above the normal level of other cars and offers the comfort advantages of a luxury car. At the same time, environmental organisations have been advocating the purchase of new cars with low fuel consumption and pushing for a maximum speed limit in Germany since the 1980s.

Symbolic Meaning of the Car

This fact shows that the car cannot be understood solely from the point of view of its practical usefulness. Rather, its construction and features are inscribed with essential

The original version of this chapter was revised. A correction to this chapter is available at https://doi.org/10.1007/978-3-658-29760-2_8

W. Ruppert (✉)
Center for Cultural History Studies, Berlin University of the Arts, Berlin, Germany
e-mail: wolfru@t-online.de

cultural and symbolic characteristics, which continue to exert a largely subconscious influence today. If one is trying to contain the harmful consequences of car culture or to eliminate them through a new conception of this object, a prerequisite for the success of all such efforts is to be aware of these symbolic meanings. They have to be incorporated into the concepts for the further development of the car.

The conversion of the powertrain to an electric motor is a sensible goal for the first step towards electromobility. However, it is by no means certain that the difficulties that have not yet been surmounted can be surmounted through new technical solutions. In the first phase of the electric car from around 1900 to the early 1920s, it was not possible to eliminate the disadvantage of restricted range, due to the limited capacity of the battery. At that time, Germany was one of the leading nations in electrical engineering and the construction of electric motors.

In the past decade, great efforts have again been made to find better technical solutions for the energy storage of electric powertrains. The development of new materials for lighter chassis and body construction is also an additional, promising path. However, all historical experience shows that technical innovations often also have side effects and consequences that only become apparent after a certain delay. This is also the case in the context of electromobility.

Culturally-Specific Characteristics of Usage

Technical innovations do not obviate the need to recognize the importance of the car for Western modernity, because the electric car also remains part of the longer history of the car as an object. This consists not just of technical data but also of culturally-specific usage characteristics for the users.

The acceptance of new concepts for the car remains bound to the quality of its cultural characteristics. The worldwide success of the car as an object of industrial mass culture is evident today in economically aspiring nations such as China and the countries of Latin America. Like no other object, the car demonstrates the value of the cultural and symbolic qualities which are inscribed in its construction and design for modern, global civilisation. In addition, when working on a meaningful new conception of this object, and even looking beyond electric mobility, it is indispensable to take into consideration the gains achieved so far for mankind's scope of action, for our "dominion over space and time".

Gains Achieved for Mankind's Scope of Action

The cultural conception of the car's use potential and its symbolic connotations is central. It is decisive for the success or failure of a new design and needs to be reconsidered with its environmental impact in mind.

Ecological Modernity

These are open questions at this stage. The necessary reformulation of modernity, which is geared towards an ecological modernity in the context of our experience of climate change, can be regarded as a creative project in the service of human civilization. To accomplish this, it is essential to be able to classify and recognize our present situation in the long-term context of its origins. This is the only way to develop meaningful concepts and convincing technical innovations for the future. The professional designers in the offices and the "decision-makers" in companies and in politics need a cultural-historical background knowledge for this, providing them with a depth of focus in their judgements.

The following text is a brief history of the car as an object, analyzing its characteristics in light of their potential for human action. The complexity of this history can only be understood as part of the cultural history of modernity.[1]

The Two Poles of the Man-Machine Relationship

This commentary from 1906 by early car driver Otto Julius Bierbaum reads like a clairvoyant prognosis:

> We are at the very beginning of its development. But it is already perfectly clear how enormous its prospects are. What we experienced with the bicycle will be repeated with the automobile—only to a greater extent. The rhythm and intensity of traffic will adapt to this almost ideal means of transport. It will begin a new era of travel, and not just for very rich people [...]. (Bierbaum 1906: 321)

Individual Mobility as a Basic Human Need

It is a fact that, since its invention in 1886, the car has experienced increasing acceptance in industrial societies. It represents in an exemplary fashion the possibilities and limits of modern technology and modern comfort in satisfying a basic

---

[1] This text was originally published in the volume *Fahrrad, Auto, Fernsehschrank. Zur Kulturgeschichte der Alltagsdinge*, edited by Wolfgang Ruppert (1993), under the title "Das Auto. Herrschaft über Raum und Zeit". It was conceived as part of a new cultural history. For the present version, it had to be shortened, with a focus on the essential points.

human need, namely that of individual mobility. The car has therefore rightly been called the leading fossil of our time.

The automobile has been given a great deal of attention over the course of its hundred-year history. But a considerable part of the literature, with its illustrations and technical details, merely meets the needs of enthusiasts and reproduces the technical myth associated with the car. Monographs such as those by Eric Schumann (1981), Wolfgang Sachs (1990) or Wolf Dieter Lützen (1986) have so far been the exception in this regard. The aim of the present text is to bring out the structural and cultural continuities of the automobile as well as its transformations.

From the beginning, cars were designed for different, everyday purposes and produced in a considerable variety of types. In Meyer's *Großes Konversationslexikon* from 1909 one finds the following entry under the keyword 'Motorwagen' (Fig. 2.1):

Motor-driven vehicle, in the narrower sense a motor-driven vehicle with no need for rails. According to the type of motor power, a distinction is made between gasoline cars, steam cars and electric cars; according to the type of car: steam calash, steam coach, steam bus, etc. Duc, Coupé, Phaethon, Tonneau, Landaulette, delivery van and truck. The most widespread and technically most accomplished are the gasoline cars. (Meyers 1909)

Fig. 2.1 The cars are still carriages, but equipped with motors. The manufacturer Lutzmann from Dessau presents various models in a motorcade, 1897. (Source: Landesbildstelle Berlin)

In order to explain the continuing quantitative expansion of the car from 1886 to the present, it is necessary to start from both poles of the man-machine relationship: the car, a machine-object, and the person who acquires it. The one pole, the vehicle, is defined as a technical-functional mechanism that provides the most original and important practical use value, namely faster movement in urban traffic, for travelling or for freight transport. The driver in no way acts as an abstract or even isolated figure: as a subject, he is integrated into the context of the cultural forms of expression of his day. His possibilities of using the car range from its use as a simple means of mobility to the spoiler-equipped "speedster", with which both the desire for mobile presence, for performance-oriented driving and the sensual experience of speed as well as—in the mode of sport—individual aggressiveness can be acted out. The elegance of expensive makes of car with aesthetically impressive and chrome-plated bodies, which serve the desire for conspicuousness and individuality of their wealthy owners, should also be mentioned.

The driver uses the car in the context of social rules, social conditions and cultural patterns. In the usual everyday forms of the acquisition of the car and the ways of using and dealing with it, a considerable role is played by socially communicated desires and conceptions of aesthetics, by ideas and information typical of the period in question. These are not re-invented on an individual basis, but rather taken up in a communicative context and as a rule merely varied.

The Car Considered as the Driver's Communicative Behaviour

It therefore seems appropriate not to explain the driver's actions solely through his relationship with the object, the car. Rather, he communicates into the public space of the street in which he is being perceived and where his audience—the other road users—is also on the move. The ways a driver uses his car can thus be interpreted as his communicative behaviour, even if there is a tension between typical group attitudes and varying degrees of individualisation.

If one looks for the underlying reasons for the fascination surrounding the car, which—despite its well-known negative effects—continues its success story even today, one must turn one's attention to the characteristics and practices that characterize the object itself in its material structure, regardless of the differences between sexes and social classes when it comes to ways of making use of the car.[2]

---

[2] By 1900, the car was already being driven by upper-class women, although light and elegant vehicles were considered appropriate for them. On gender-specific uses of the car, cf. for example Steffen (1990: 133 f.).

The Car as an Object of Industrial Culture

Unlike the railway, the car was not restricted to a rail network, but was in principle able to operate independently, in open terrain. It soon became apparent, however, that the increasing speed of the car caused enormous trails of dust to be thrown up and that the roads that had been standard up to then were no longer adequate, so that the practical value of the car only became apparent in connection with an improved infrastructure (cf. Merki 2002; Möser 2002). This led to the expansion of a suitable road network based on the principle of a smooth flow of traffic, a project that has been pursued ever since and has dominated conceptions of modernisation (cf. e.g. Schmucki 2001).

Decision to Prioritize Direct Movement Through Urban and Rural Spaces

Accordingly, cities were redesigned in the course of the twentieth century and the landscape was further transformed.[3] In the culture of modernity, a—barely considered—decision in favor of direct movement through urban and rural spaces was made—placing it above the preservation of nature and other human needs. With the construction of motorways, which cut through the landscape in the form of straight raceways, the intensive networking of the regions began in 1933, for example with the north-south route from the Baltic Sea via Berlin-Nuremberg-Munich to the Alps (Stommer 1984). Their routing was designed as an engineering concept in such a way that the natural obstacles were smoothed out, with river valleys crossed by means of wide-span bridge structures.

Since the 1950s, massive urban motorways have also been built inside cities as quasi-natural borders between the city districts. They separated different traffic and neighbourhood spaces from each other.

In urban space itself, the car was granted a high priority. Vehicles parked on both sides of the street have been obstructing the roads ever since the mass motorisation that started in the 1950s. Parking spaces were created in traffic areas that were still vacant, which mostly only brought temporary relief. New forms of architecture were designed to meet the specific needs of parking and protecting cars: garages for single-family houses, underground garages for residential complexes

---

[3] Car culture is discussed as the most important form of everyday culture by Fünfschilling/ Huber (1985) and by Bode et al. (1986) in their catalogue accompanying an exhibition in Munich.

and supermarkets, multi-storey car parks for inner-city traffic congestion zones (cf. Honnef 1972). In order to regulate traffic, traffic signs increasingly served as a visual expression of traffic regulations. Traffic lights began to regulate the rhythm of movement of the cars in the traffic flow on the basis of time sequences. Compared to the free street space around 1900, a hegemony gradually developed, in which the artefacts of automobile traffic and the dominion of the structural order derived from it were set against weaker road users, such as cyclists and pedestrians or children and wheelchair users. Furthermore, the operation of the car required a network of services which expanded into a separate economic sector based on the division of labor. The gasoline pumps of the first decades became service stations offering a wide range of services from car washing to tyre changing (see Polster 1982). A longer overview of the numerous material objectivations of car culture would cover the following: car service, car dealerships, car centers, car repairs, car accessories, car boutiques, car tuning, car radios, car phones, car sport, sports cars, car comfort, luxury limousines, nippy runabouts, spoilers, car styling.

Hegemony of the Artifacts

The car and its systemic requisites—such as the road network, traffic architecture and services—developed into a characteristic element of industrial culture.[4] This can be explained above all by the far-reaching democratization of the car as a privately used vehicle.

Democratization of the Car

Mass Motorization

In 1907 automobiles began to be counted. The statistics provide a picture of the course of motorization in the German Reich and the quantitative diffusion of the car (cf. Krämer-Badoni et al. 1971: 11, 16). They show an astonishing continuity and at the same time a clear dependence on political and economic history. While in 1907 the number of automobiles was estimated at around 10,000, one can assume that in the first two decades of automotive history cars were considerably less

---

[4]A source on the state of development in the 1920s cf. Allmers et al. (1928).

prevalent. The quantities built were used, for example, for models such as the *Daimler Belt Trolley* from 1895 with approx. 130 units or for the *Benz Ideal* from 1901 with about 300 units. In this early period, however, there were about twice as many motorcycles, which were regarded as "vehicles for snobs".

Dependence on Political and Economic History

Only after the inflation and currency reform of 1923 (98,000) and the ensuing economic consolidation did the number of cars rise to 261,000 by 1927, to 422,000 by 1929 during the "heyday" of the Weimar Republic, and to 510,000 in 1931. Due to the world economic crisis, the number fell slightly in 1932 to 486,000, only to rise again to half a million in 1933 and triple again during the Nazi era up to the beginning of the war: in 1935 there were already 795,000 cars; in 1937 there were 1,108,000 and 1,426,000 in 1939. After the war began in 1939, the use of private cars was suspended. However, this break in private motorisation was quickly made up for in the post-war years.

Fig. 2.2 The long-distance journey from Paris to Berlin ended at the Berlin-Westend trotting track in 1901. (Source: Landesbildstelle Berlin)

The Motorized Society

Between 1951 and 1960 the number of passenger cars increased tenfold. Around 1954/1955 the number of automobiles in the now much smaller area of the Federal Republic of Germany reattained the highest pre-war level, with about 1.5 million vehicles. In his analysis of the significance of the car for mass mobility, historian Peter Borscheid (1988: 122) assesses the time around 1960 as marking the "epochal boundary" to motorized society, especially since this trend continued in the 1960s: 1963: 6,807,000; 1971: 14,377,000 cars.

Increase in Real Income

This quantitative process was associated with a qualitative historical development that was embedded in a context of specific socio-historical conditions. Mass motorization was only possible on the basis of the increase in the real incomes of broader strata of the working population, which took place after 1957/1958 in an historically revolutionary manner for both workers and white-collar employees. Their need for increased spatial mobility found an attractive cultural form in the new opportunities for activity offered by this industrial object, namely as motorists.

The car brought considerable changes to the scope for decision-making in the search for work. Regional mobility and internal migration in the Federal Republic increased overall. The growing volume of traffic in turn functioned as a "practical imperative" in the development and sealing of roads (cf. Linder et al. 1975). Better road surfaces reduced both driver stress and wear and tear on cars.

The daily distance people travelled by car or public transport increased. Whereas in 1960 the average distance was 12.5 km, in 1970 this figure had risen to 20.6 km, with an upward trend to 40 km to date (cf. DIW 2011). The faster flow of traffic along the road network, which is still being expanded today, extended the travel radius for those commuting between their place of residence and their workplace.

Improvement in the Standard of Living

In addition to the manifest usefulness of the car for commuters, the purchase of a car also came to symbolize the increase in the standard of living, which went hand in hand with the extension of personal freedom of movement to the place of residence, manifested in the form of the Sunday excursion.

As a result, the difference between the city and the surrounding countryside diminished, a new form of urbanisation of the countryside began, and the objects of industrial mass culture increasingly filled a landscape that had previously been preserved.

Despite continuous motorization, however, full motorization was not achieved even in the 1970s and 1980s. There are now around 43 million passenger cars in Germany. For a minority of the population, the car remains unaffordable for financial reasons.

Car Owners

In the first decades, the automobile remained a pure luxury object, owned and driven by estate owners, people of private means and bourgeois professionals such as doctors. For the former group, the car was partly a machine for sport and pleasure, partly an object that met the requirements of representing status (Fig. 2.2). An advertisement by Benz und Co. in 1888 for the "Patent-Motorwagen" illustrates the advantages that were supposed to make a purchase interesting for people at the time (quoted from Sachs 1990: 14). The vehicle was not only "comfortable and absolutely safe", but "always immediately operational". Due to the "gas operation by petroleum, gasoline, naphtha etc." it was classifiable as a "complete replacement for horse-drawn wagons". In addition, it was cheap, with 'very low operating costs', since it 'saves the cost of the coachman, the expensive fittings, maintenance and upkeep of the horses'. The target audience for this advertisement and with whom the story of the procurement of cars began, were those wealthy private individuals whose standard of living naturally included a carriage with horses. However, the "convenient" operational capability claimed in the advertisements was by no means entirely a given. In contrast to the time-consuming hitching-up of horses, the automobile could be started without delay, but the machines themselves were not yet fully developed and thus prone to failure. Therefore, in place of a coachman, it was necessary to employ a chauffeur, who—doubling as a mechanic—was also able to service, repair and keep the car ready to drive (Picture 2.3).

In the 1920s, other sections of the bourgeoisie and businessmen began to use the car to reshape their lives. In addition, simple utility vehicles such as cycle cars were also built, but the number of units remained low. For example, between 1924 and 1925 about 1500 units of the *Mollmobil* were manufactured, a light vehicle based on the technology of bicycle construction and which today seems primitive. With a simple box chassis, it reached a maximum speed of 35 km/h.

Picture 2.3 In the first half of the twentieth century, horse and cart, tram and the "modern" car still moved as equals side by side in the street (1921). (Source: Landesbildstelle Berlin)

Price Reductions as an Industrial Mass Product

Only in the third decade of the twentieth century did the possibilities for purchasing a car increase, in several stages, above all due to price reductions resulting from manufacturing methods that turned the car into an industrial mass product.

Towards the end of the 1920s, a cheap, mass-produced car "for all" moved into the realm of the technically possible. This long-standing and latent dream of the middle and lower classes was taken up by Nazi propaganda, which presented the "Volkswagen" as an acquisition within reach of every saver. The utopia of the motorized national community seemed to be the expression of a consistent historical trend: not just the wealthy but also workers and manual laborers (*Arbeiter der Faust*) were meant to benefit from the technology. The Nazi leisure organization "Kraft durch Freude" combined the development of the *KdF Auto 22* with an advertising campaign in which the Führer, Adolf Hitler, was portrayed as the initiator of future achievements.

Goebbels' speech at the opening of the International Automobile and Motorcycle Exhibition in Berlin in 1939 emphasized the connection between large-scale production and the expansion of the sales markets through the expansionist NS foreign policy, as well as the price reductions brought about by standardized mass production, using the *Volksempfänger* (people's radio receiver) as an example:

> For example, the customer base for radio sets in today's Reich has grown so large that we are now in a position to significantly reduce production costs, thanks to the mass consumption that is already guaranteed internally. The same also applies to the production of German films etc. However, the car will only be able to compete on a global scale in terms of pricing if the possibility of large-scale series production is ensured. This presupposes in all circumstances an adequate economic sphere of our own. (Goebbels 1939: 15)

As from 1938, an instalment savings plan had been on offer, which promised the acquisition of the object of desire, namely the Volkswagen, and payments were made into the accounts of a total of 336,000 savers. However, their expectations were disappointed since in the end production did not get under way at the newly built Volkswagen plant in Wolfsburg (cf. Mommsen 1996).

Until the 1950s, the purchase of a prestigious and comfortable car was reserved exclusively for the upper classes. Motor vehicles for workers and white-collar employees were initially mainly restricted to motorcycles or scooters. In the early 1950s, the purchase of a moped had become possible even with a very modest income. Compared to two-wheeled vehicles, which offered no protection against bad weather and the cold, the multi-wheeled vehicle types with an enclosed driving space were obviously perceived as a gain in comfort: for several years, small cars like *Lloyd, Goggomobile, Maico, BMW-Isetta* or the *Heinkel Cabin Scooter* reflected the trend of symbolizing the increase in mass purchasing power.

Sold in large-scale series in the 1950s, the *VW Beetle* offered the advantage over these small cars of a reliable family vehicle with considerably more freedom of movement for passengers (cf. Hickethier et al. 1974).

Mass motorization in the 1950s and in the wake of the "Economic Miracle" in the 1960s can be viewed as a phenomenon of great social and cultural significance. Of 100 private households, 27 already owned a passenger car in 1962; by 1973 the number had risen to 55. For people living in the countryside, the mobility gained by buying a car brought urban spaces within reach and made the short trips necessary to everyday life considerably easier. In addition, driving to Italy for a holiday, using one's own car—camping on the Adriatic coast, for example—soon became one of the attainable goals and improvements in the collective standard of living.

Industrial Mass Product

In the course of its quantitative proliferation, the car developed into an important paradigm of industrial mass culture, in which its dual nature became visible. On the one hand, the history of the automobile was associated with the history of an important and growing branch of industry, the automobile industry, in which innovative rationalisation processes with exemplary character for the mass production of goods were developed. The profit expectations associated with the production of large-scale series resulted in an active marketing policy on the part of the companies. Through advertising and vehicle design, market strategies were invented to address potential buyers and direct their wishes specifically to the manufacturers' product range. On the other hand, car buyers were focused on the fulfilment of the promises of practical usefulness conveyed by advertising, and the ideals associated with it. They played a part in the socially communicated concepts of modernity and their visual codes.

The forms of production reflect the history of industrialisation in the twentieth century. In the first decades, manual manufacturing processes predominated in the production of small-scale series. In the case of models that sold well, the assembly technique was the industry standard form of individual assembly stations placed alongside each other. The parts were manufactured separately then fitted to the car by hand.

Introduction of the Assembly Line

In Germany, too, engineers had since 1900 been looking for ways to raise standards of rationality and, in the context of business and study trips, began to test the transferability of the more advanced American methods. With the successful introduction of the assembly line in 1913 by Henry Ford, a new model for industrial mass production became established in the assembly technology of automobile construction. On the basis of precise measuring methods and the Taylorist division of labor, the various production steps were allotted to fixed individual workstations on a moving conveyor belt. While the vehicle being assembled was at his workplace, each worker had to perform the pre-planned operations in a precisely measured time, before the vehicle moved on. The pace of the belt dictated the working speed, to which the individual had to adapt. The standardized parts were mass-produced in other production units and delivered for assembly, which facilitated a considerable increase in the effectiveness of the assembly process.

From 1914 onwards, this flow production made it possible to reduce the price of the "Model T" (which was produced between 1908 and 1927) in an exemplary fashion. Whereas in 1909 it had cost 950 dollars, by 1917 its price had fallen to 350 dollars, and in 1923 to 290 dollars, thanks to the assembly line. At the same time, the number of buyers increased due to the price reduction. The model *Tin Lizzie* sold a total of 15,007,033 units.

Because of the smaller national market in Germany and a larger number of competing manufacturers, but also because of the economic crises and inflation after the First World War, it took until 1924 for the first car to be produced on an assembly line: With the so-called *Tree Frog* the car manufacturer Opel—borrowing from the French model of the Citroën CV 5—reached wider groups of buyers, mainly from the middle classes, who now made their entry into motorisation.

Product Policy

In the 1920s, the importance of a product policy that took into account the changing tastes of the times and the internal differentiation of the model range as designable factors grew. In competition with Ford, whose "Model T" began to look old-fashioned in the mid-1920s, General Motors had developed a new product strategy, which involved the gradation of its automobiles in a hierarchy from expensive to inexpensive, so that for upwardly mobile people, with increased income and seeking greater prestige, it was possible to buy a higher-quality vehicle from the same manufacturer. This concept was called "Sloanism", named after the then director of General Motors, Alfred Sloan, (cf. Sachs 1990: 98 f.).

Increasing Attractiveness

A further significant step in the development of an industrial mass product was taken with the changeover to a model policy based on changing fashions and, later, the planned obsolescence of the vehicles. The more rapid change of car models was intended to expand market share, but also to prevent saturation of the market (by 1927 Ford's "Model T" had become completely unsaleable). The calculated increase in the attractiveness of the models, which were also technically modified, made the work on the shape of the car, the "styling", which was supposed to make the body of the car more attractive, an important factor in company policy.

Brands

While the secret of the Volkswagen's success after the Second World War was based on its large-scale series with only minor improvements to the model and a comparatively low price, Opel endeavored to apply the tried and tested concept of its American parent company, General Motors. Along with its car for the middle classes, the *Record*, and the large, expensive prestige model, the *Captain*, Opel successfully launched the *Cadet* in the 1960s, a car designed to stand out from the mass-market *Beetle* by offering higher prestige values. The *Cadet* was meant to appeal to upwardly mobile people and at the same time bind them to the brand.

Individual Distinction

This became the general trend. Cars that stood out from the masses in terms of their prestige value gained in importance. As motorization progressed in the 1950s, buyers' needs for individual distinction found a medium in the choice of car model and, in the optional extras, a rich field of expression for cultural codes, which enable individuality to be constituted in and through an object. The product policy of the car manufacturers pursued the strategy of translating the wishes of consumers for social differentiation into a model range of different types of cars. This proved to be an increasingly important cultural function of the car.

Cultural Function of the Car

The Car as an Object of Modernity

The French structuralist Roland Barthes devised the image that today cars are "the exact equivalent of the great Gothic cathedrals" of the Middle Ages (Barthes 1964: 88[5]). He based this comparison on three reference levels: "I mean the supreme creation of an era, conceived with passion by unknown artists and consumed in image if not usage by a whole population which appropriates them as a purely

---

[5] Roland Barthes, "The New Citroën", in *Mythologies*, trans. Annette Lavers (New York: The Noonday Press, 1991), p. 88.

magical object" (ibid.). What occasioned Barthes to speak of this "supreme creation of an era" was the presentation of the Citroën DS19 in 1955.[6]

To the extent that, in the course of the twentieth century, the emotional and symbolic needs of considerable sections of the population were linked to the car, the question remains as to what characteristics made this machine an expression of a specific experience of modern times.

An indication of the cultural meanings that the car took on even in its early days can be found in the way in which artists thematized and processed the phenomena associated with it as pictorial subjects. In the eyes of the Italian futurists, the car—along with the aeroplane—became a symbol of the new dynamic age because it represented the trends of innovation and the spirit of technical civilization.

Futuristic Manifesto: Beauty of Speed

In the Futurist Manifesto of 1909, the machine that generated speed was aestheticized: "We affirm that the world's magnificence has been enriched by a new beauty: the beauty of speed. A racing car whose hood is adorned with great pipes, like serpents of explosive breath—a roaring car that seems to ride on grapeshot is more beautiful than the Victory of Samothrace".[7] Super-charged with meaning, the myth of technology and progress was here imbued with the celebration of dynamic movement: the car was stylized as a cult object of speed.

Practical Values

In 1906, an anonymous commentator made an astute observation on an essential characteristic of modernity, namely the compression of the factor of time, which was subjectively expressed in the needs of modern man: "One wants to be transported as far as possible in the shortest possible time, and our fast-paced generation has provided itself with new means of transport" (quoted by Sachs 1990: 191).

---

[6] Later, the car, which was elevated to cult status, was also promoted as a goddess because of the linguistic similarity of (la) déesse(= the goddess) with the letter DS.

[7] F. T. Marinetti, "The Founding and Manifesto of Futurism", trans. R.W. Flint, in Apollonio, Umbro, ed. *Documents of 20th Century Art: Futurist Manifestos*, New York: Viking Press, 1973. pp. 19–24.

Exponentiation of Power

The car is a "magical object" (Barthes) that emerged from this resolve. It was built as a machine for more rapid movement, whose capacity for linear acceleration far outstripped the natural possibilities of the body. Measuring this power in terms of horses and thus "horsepower" (hp), which had previously been chiefly used in individual transport, is a reference to the original perception of the exponentiation of power that the car afforded.

An ethnological author interested in the "culture of everyday life", Michael Haberlandt, had already praised "the great increase in individual mobility" with regard to the bicycle around 1900, and which he described as a restructuring of traffic:

> It brings into the immense, countless meshes of collectivist traffic the unbound circulation of individuals who enliven the desolate intermediate streets and side streets with fleeting swarms of people, everywhere criss-crossing each other's paths, filling the broad meshes of mass traffic and creating movement everywhere where calm and immobility used to prevail. The emancipation of the individual from the cumbersome public transport system thanks to the bicycle, the new freedom of movement granted to the person [...] is a cultural advance of immense significance. (Haberland 1900: 127 f.)[8]

The car continued the civilizing trend that had already been established by the bicycle. A more abstract dimension of practical use values, which gave the "supreme creation of an era" (Barthes), the car, its deeper meaning, was surprisingly precisely formulated in the "Allgemeine Automobil-Zeitung" in 1906:

> The car wants to impose man's dominion over space and time, by means of the speed of movement. The entire enormous apparatus of the railway, railway network, railway stations, signal stations, monitoring services and administration now becomes irrelevant, and man rules over space and time in comparative freedom. (quoted by Sachs 1990: 19)

---

[8]The historian Hans-Erhard Lessing (2003) also sees what the bicycle and automobile have in common in the principle of self-mobility. In order to make it clear that the origins of self-determined mechanical mobility lie with the bicycle, he accordingly titled his history of the bicycle "automobility".

The Car as an Objectivation of the Modern Quest for Speed and Locomotion

If one follows this analysis, the car is to be seen as an objectivation of the modern striving for "speed of locomotion", from which the "dominion over space and time" arises. Given this goal, the history of the car as an object is to be viewed as a process of optimizing a corresponding technical potential. "Speed of movement" and the "unrestricted circulation of individuals" have become guiding models of industrial civilization.

In their evaluation of a basic characteristic, the commentators on this process of modernization were in agreement: the "motor vehicle" had made individual, self-determined freedom of movement possible, from place to place, in both urban and rural areas. Ever since the early years, this new freedom of movement has been regarded as a compelling practical value and as an experience specific to travelling by car.

Freedom of Movement

With the help of the car, the individual experiences himself as no longer confined to acting collectively: he now has individually determined rhythms of time, moves at an individually determined speed and is able to take individually determined pauses. The comfort and convenience thereby made available in everyday culture also contributed to far-reaching changes in ideas about the city and the order of the modern world.

With reference to Barthes' thesis of the "magical object", however, further questions remain: is the magic of the car based on the potential of the technical apparatus to expand the natural capacities of the human body and to "erase" local restrictedness?

In any case, by driving a car, the motorist—within the limits of the man-machine movement and its functions—can experience himself as an active subject and "create" himself as an individual in and through freedom of movement.[9] In this sense, the car was designed in terms of an "I prosthesis" (Wolfgang Sachs) and a "prosthesis of our mobility" (Borscheid 1988: 135).

---

[9] Richard Sennett (1991) refers to the modernity of this belief in self-creation.

The Cultural Configuration in the Context of Modernity

Motif of Movement as the Motif of the Time

Of course, the car is not the only new means of transport that facilitates rapid movement. Its emergence must be seen in the context of a social and cultural configuration in which mobility and speed became increasingly charged with meaning and an essential factor in the general framework of relations with modern culture. In 1914, an outstanding architect such as Walter Gropius, who was extremely perceptive when it came to civilizational developments, listed the following means of transport as "symbols of speed": "the automobile and the railway, steamship and sailing yacht, airship and airplane". They are embodiments of "the problem of traffic movement". Gropius clairvoyantly declared the "motif of movement" to be the "decisive motif of the time" (Gropius 1914: 32).

This assessment found a counterpart in the history of the experiences of the modern individual. Around 1900, the cultural philosopher Georg Simmel, in his fundamental 1903 essay "Die Großstädte und das Geistesleben" (*The Metropolis and Mental Life*), referred to the unprecedented urban individuality resulting from rapid movement:

> The psychological foundation, upon which the metropolitan individuality is erected, is the intensification of emotional life due to the swift and continuous shift of external and internal stimuli.[10]

Kinetic Revolution

This "swift and continuous shift" had in fact been described as early as the middle of the nineteenth century as a core element of modern culture. The characteristics of the car, which served to satisfy the individual need for more rapid mobility, represented the transient and fleeting aspects of modern life. The car is embedded as a movement machine in this more comprehensive cultural-historical process of the kinetic revolution, which is objectified in the forms of material culture.

---

[10] Georg Simmel, "The Metropolis and Mental Life" (1903), in Gary Bridge and Sophie Watson, eds. *The Blackwell City Reader*, Oxford and Malden, MA: Wiley-Blackwell, 2002, p. 11.

In the course of the nineteenth century, the acceleration of movement led to the invention of an ensemble of new means of transport. These included trains, steam cars, bicycles, automobiles, motorcycles, airplanes. The opening of the first railway line between Nuremberg and Fürth in 1835 can be regarded as a symbolic event in Germany. In addition to the railway, the still cumbersome "steam car", built for road operation, was one of the first machines of the new mobility. As a reaction to the increasing distances within the city that the process of urbanisation entailed, rail lines were set up on which horse-drawn buses travelled until they were temporarily superseded by steam engines and finally replaced by electricity-driven trams. Industrialization—the innovative competence of mechanical engineering in the development of new mobility machines—and the increased social need for mobility were mutually dependent. With the railway, an initial form of increased acceleration and the intensification of movement had developed, which, however, was bound to the fixed lines of the railway network and, in addition, remained a collective form of travel. The need for the individualization of mobility eventually created the new generation of machines and ushered in the phase of individual transport. The increasing suitability of the bicycle brought a boost in individualization for mass transport. Furthermore, in the last decades of the nineteenth century there were various two-wheeled motor vehicles with gas engines or steam engines. However, these were not produced in large numbers, since their practical value remained restricted due to technical unreliability and also due to the fact that the road network had been extended only to a limited degree. It was only after 1900, but especially during and after the 1920s, that the motorbike with a gasoline engine became more common, when it became affordable for workers and white-collar employees due to the price reduction brought about by mass production.

Individualization of Mobility

The car represents a second push towards individualization. It is to be interpreted as a continuation and supplement of this fundamental trend of cultural modernity. The individual buyer of the "motor wagon" was able to acquire the potential for mobility as an instrument for dissolving the boundaries of space and the local bonds of individual life, in a material form—as a machine. The new forms of cultural experience associated with the car constitute a break in the history of civilization and mankind, the scope of which cannot be overestimated. This assessment becomes all the more striking when one considers that even in the eighteenth century 80–90% of the population did not travel any further than a day's journey from their place of residence in the course of their entire lives—partly because this was not an aim, due to their embeddedness in the estates-based society, partly because the constraints of everyday life did not permit greater latitude for action. With the emergence of the ensemble of new forms of transport in the nineteenth century, the cultural experience of modernity intensified as a result of this increased mobility: the increase in the speed of movement in everyday life and the rationalization of bourgeois professional life produced the phenomenon of haste. Edginess also arose as a result of the tension between the ever more precise scheduling requirements of business activities and the hindrances caused by traffic.

Dissolving the Boundaries of Space: A Break in the Civilizational History of Mankind

The Panoramic View of Space

The associated over-excitation and mental overload led to false reactions and catastrophes, as documented by an increasing number of accidents. At the same time, with the increasing speed of the railway as a means of transport, a new form of modern perception had emerged: the panoramic view of space (cf. Schivelbusch 1977).

This was linked to the degree of speed. The faster movement caused a growing fleetingness of connectedness to a locality and loosened the contact to the landscape through which one was travelling, the experience of approaching and moving away from a place alternated more frequently (cf. Kaschuba 2004). Objects seemed to fly past in nearby space, while the landscape in the distance passed by more slowly. This aesthetic experience also applied to cyclists, as Haberland observed:

> The fleeting glimpse, the swiftness of the changes, the ephemeral images garnered in the course of a rapid transit on two wheels, while the eye constantly monitors the route, this new way of looking, in a sense also requires a new aesthetic. (Haberland 1900: 130)

While the speed of the railway passenger remained tied to the collective movement of the train, the cyclist was able to adapt, intensify or slow down the input of "ephemeral images" (*Momentbilder*) according to his individual needs by adjusting and directing his movement. Following on from and consistent with this experience of modernity, panoramic perception in conjunction with individual control over speed also became an internalized form of experience for the car driver.

With the increase in the number of road users, the encounters between drivers on the roads in turn became increasingly anonymous.

Isolation While Driving

These encounters became more abstract, reduced to the objective traffic rules. Travelling separately in one's own vehicle meant emancipation from the collectivity of the railway or bus but at the same time isolation while driving. It was only with passengers in the car that a small community came into being, in the shared movement through space.

Symbolic and Aesthetic Aspects

The conviction that the "automobile is a symbol of progress" (Bierbaum 1906: 325) was already shared by its supporters in 1906. From the outset, this symbolic charge was associated with the social and cultural uses of the car, which would therefore fulfill Clifford Geertz's definition of the term symbol: "any object [...] which serves as a vehicle for a conception, the conception [being] the symbol's meaning."[11]

Aestheticization of the Car

Such symbolic meanings were objectified partly in the equipment and form of the car itself, in its design, and partly in the connection with the contemporary myths of progress produced in advertisements or photographs. In addition, being simultaneously an actor and a consumer, the motorist was able to add to the object individual forms of symbolisation that approximated his needs. These included symbolisations of social status, individual taste and the desire for distinction, but also the ownership of the car as a symbol of power or freedom. Corresponding aestheticizations of cars and the images that promoted them became part of the commercial production of attractiveness in industrial mass culture in order to attract buyers. The "magical fascination" of the object, as Barthes noted, attained and still retains

---

[11] Clifford Geertz, *The Interpretation of Cultures: Selected Essays*, New York, Perseus Books, 1973, p. 89.

such an intensity precisely because of these direct and indirect inducements based on aesthetic and cultural codes.

Aesthetic and Cultural Codes

The Myth of Technology and Speed

When the history of the "motor car" began, there were poorly conceived inventions that competed with each other. Work was carried out in parallel on improvements to the gas engine, the electric motor and the gasoline engine in order to develop a suitable powertrain. There was also competition between two-wheeled, three-wheeled and four-wheeled carriages with gasoline engines.

Electric Cars

As early as 1885 Carl Benz had tested a motor tricycle that reached a speed of 10–15 km/h. His four-wheeled carriage with a gasoline engine from 1886 differed only slightly from this. In the first decades, the electric car still played a role in cities, competing with the gasoline-powered vehicle. It was well suited as a taxi-cab and as a transport vehicle and experienced a temporary comeback after the First World War. However, since the problem of the heavy battery and charging it could not be solved satisfactorily, the electric car's importance declined (cf. Mom 2004).

Car Races

As their power and performance increased, the new machines were coded with the myths of technology and the "dominion over space and time" typical of the period. Early in the history of the object, a ritualized form of testing the relationship between man and machine emerged in the form of car racing. In 1895, the first race took place on the Paris-Bordeaux-Paris circuit, and in 1898 a race was held in Germany from Berlin to Potsdam and back, a circuit on which the winners reached an average speed of 25.6 km/h. In addition to testing the driving characteristics and comparing the technical performance of the car, the principle of continuously increasing the speed was already at the forefront of these early races (Eichberg 1987: 162). The benchmark was based on a concept of performance that exercised increasing fascination as a symbol of the guiding principles of modernism. The measurement of speed and the increase in speed records constituted an implementation of the scientific apparatus of technical modernity, which had become an essential factor in the concept of progress in the nineteenth century (cf. Borscheid 2004).

Due to these high levels of symbolic meaning and the yardstick of performance optimization inherent in it, car manufacturers developed a special interest in these car races. The aspect of extreme stress at "full speed" was foregrounded in Meyer's *Konversationslexikon* of 1909, in a definition of the racing car that remains valid today:

> An automobile capable of great speed. Since in such a vehicle every screw, every bolt, in short the smallest detail is stressed to the extreme when the car is being driven at full speed, the racing car serves as a touchstone for the reliability of the construction and quality of the materials. (Meyers 1909: 191)

The speed capability increased astonishingly quickly. In 1911, the "Benz Lightning" reached a world speed record of 228.1 km/h, which held until 1924. For a touring car in 1909, an average speed of about 80 km/h was considered standard for a machine with 150 hp.

As speeds increased, car races turned into a sensual experience, with an intensification of acoustic and visual stimuli, of experience and pleasure in the amassing of perceptions.[12] As early as 1906, Bierbaum observed a peculiar tension at the initial races: "The work of competition becomes a public drama" (ibid.: 320). This was equally true for drivers and spectators.

After a car test track, the AVUS, was completed in Berlin in 1921, the races in the 1920s and 1930s developed into the highlights of an unparalleled technical cult. The myth of technology found a fitting and timely expression in the sports coverage of the races. As Barthes put it, myth is "a mode of signification, a form…it is not defined by the object of its message, but by the way in which it utters this message."[13] In the constant process of presenting the "new", the promise of happiness and the image of controllable progress were conjured up and reproduced in industrial culture. With perpetual technical innovation as a fetish of the history of objects, the car played a part in this myth and intensified it (Fig. 2.4).

---

[12] With the age of television broadcasts, the sensual dimension of the many stimuli intensified, since the "fleeting" movement of the racing cars is followed by several track cameras..

[13] Roland Barthes, "Myth Today", in *Mythologies*, trans. Annette Lavers (New York: The Noonday Press, 1991), p. 107.

Fig. 2.4 Grand International Automobile Race 1937: Racing driver Bernd Rosemeyer passes through the northern curve of the Avus. Speed, engine power and the competition between the drivers made the car races an exciting, bond-forming event of mass culture. (Source: Landesbildstelle Berlin)

Symbolization of Social Status and Prestige

Whereas the first motor vehicles were basically carriages with a rear-mounted engine, around 1900 an independent form of the car developed (Petsch 1982: 36). In touring cars, the engine was now at the front and the body formed a space for the driver and passengers. A culturally determined form of representation was created from the technically determined functional form. However, this was in keeping with older conventions, with courtly society providing the clearest example, where hierarchies of rank and prestige had been represented in the form, fittings, comfort and decoration of carriages. In 1899, the American cultural sociologist Thorstein Veblen had placed these ways of dealing with status objects in the context of an historical anthropology, formulating it thus: "The conspicuous consumption of valuable goods is a means of reputability to the gentleman of leisure" (Veblen 1986: 47).[14]

Form of Representation

Following this tradition, the car soon assumed the function of symbolizing the position of its owner in society. The glamor of a designer automobile body provided another medium to express the need for distinction of the monied aristocracy and of aristocrats by birth (Fig. 2.5). In order to offer the flair of exclusivity, the designers also used traditions and art forms which, with their connotations of stylistic elegance, also guaranteed distinction in relation to simple mass-produced products, using a

Fig. 2.5 In the Mercedes S sports car, the elongated engine block and the massive radiator demonstrated the power of the machine. A company specializing in automobile bodies equipped this one with designer elements such as the chrome exhaust pipes, which increased the prestige and made the car a medium of social distinction for the upper classes (1920s). (Source: Landesbildstelle Berlin)

---

[14]Thorsten Veblen, *The Theory of the Leisure Class*, Dover Publications Inc., Mineola NY, 1994.

wide spectrum of aesthetic forms, ranging from chrome on the radiator grille to ornamental scrollwork and white-wall tires.

Expressiveness of Form as Social Communication

The aesthetic expressiveness of the form also served the purpose of social communication, as a sign of an elevated position in society. In the course of the history of objects in the twentieth century, the car thus became a medium and symbol of socio-historical structures. Wealthy businessmen were already among the early car owners when it still possessed the aura of the new and signalled progressiveness, but on the other hand presupposed middle-class income to cover the cost of maintenance. Only in the process of mass motorization after the Second World War did white-collar employees and workers achieve a symbolic ascent in society, namely through their image as consumers.

For upwardly mobile people, the change from a small to a middle-class car—for instance, from the *Beetle* to the *Opel Record* or the *Ford Taunus*—connoted a prestige-bearing representation of individual progress. In many cases, these effects of mass motorization were later interpreted as a symbolic expression of the dissolution of class society, since in the 1970s larger cars became affordable for skilled workers with higher incomes (Beck 1986: 123). Their purchase appeared to symbolize proximity to and affiliation with other drivers and owners of the same brand of automobile from the upper class. Any skilled worker, employee or civil servant with a middle income who parked a Mercedes in front of his house or apartment was thus placing a status signal in public space and in his own residential environment.

The Car as a Signifier of Social Hierarchies

The increasingly widespread ownership of exclusive cars relativized their suitability as a sign of social class and position, but the "fine distinctions" (Bourdieu) merely shifted into the differentiations in the manufacturer's product range and the price hierarchy. For example, while the cheaper Mercedes models were only just accessible to skilled workers who earned a good living, the expensive upmarket models were still reserved for those in management and for company owners. The car thus continued to serve as a signifier of social hierarchies. Of course, the use of dark-colored limousines suggesting an aura of authority by members of the functional elite of business life was not limited to the purpose of driving to a meeting with a business partner. Rather, the symbolic-aesthetic dimension of a differentiated sign system enters into the choice of the type of automobile: the brand and the car's features must primarily express the internal position of the company representative or manager in relation to his business partners, but at the same time represent the company itself to the outside world.

The official cars of state office-holders were in any case mostly custom-built, commensurate with the aesthetic representation of state power and the dignity of the office-holder. The large Mercedes of Federal Chancellors Konrad Adenauer and Willy Brand can serve as illustrative examples.

Automobile Design and the Imagery of Civilization

The design of automobile bodies also included early visualizations that represented metaphors of progress typical of the period. I would like to use a few examples to illustrate the connection between cultural contexts and the specific meanings of the form of automobiles.

One of the first sculpture-like forms to emerge for cars with powerful engines was an elongated hood that visually illustrated the power and direction of movement of the underlying "horsepower". The so-called "boat style" developed around 1910 into a form typical of the period, which can be viewed in the context of the emphasis on technical civilization, but also against the historical background of the German Empire prior to the First World War. Modern ships were regarded as symbols of modernity due to their speed and technical perfection as well as their functional design. Therefore, the adoption of the boat form for automobile bodies is to be interpreted as bestowing on the car multiple layers of prestige: it combined the connotations of progress, rapid movement and modern technical rationality with the stylistic ideal of functional elegance and the canon of Imperial values.

Two decades later, another form of civilizational imagery emerged, namely the streamline. With increasing speed, reducing air resistance became a central criterion for improvements in the shape of the automobile. Reducing flow resistance and the speed thus gained became the starting point for the new aesthetics (cf. Burkhardt 1990: 221 ff.). As early as 1909, the body design of the world record-holding car, the *Blitzen-Benz*, was determined by aerodynamic forms. The torpedo shape and the drop shape were regarded as ideal parallels to the ultra-modern construction of the Zeppelin.

Streamlined Form

At the end of the 1920s, the streamlined form emerged, the cultural semantics of which were rooted in rapid mobility. In the 1930s, the stylistic features of the flowing lines that had developed from functionality—for example elongated mudguards—became an aesthetic form in their own right, illustrating the fascination of speed as a dimension of the myth of technology (cf. Lichtenstein/Engler 1992). The designers soon also used the streamline for immobile objects such as pencil sharpeners, radio sets or refrigerators. This form represented a promise of modernity typical of the period, but also the functional aspiration to overcome individual problems with ease in the everyday framework of an ever more advanced mass culture.

Although of European origin, the streamline was increasingly cultivated in the USA in the 1930s and finally returned in the 1950s with the American objects of mass culture in the "dream car style" as a fashion typical of the time. In this context, the streamlined, chromed "street cruisers" with shark mouths signaled the increased standard of living. In the 1950s, various models of the Opel *Record* or the *Captain* were designed with this imagery in mind.

Individual Taste and Subjective Identity

For a long time, the increase in the comfort of "modern man" remained identical with belonging to the wealthy upper class. The popular phrase "being able to afford more" was a cipher for social advancement and referred to the reality of social inequality as well as to the unequal chances of being able to afford the objects of material culture. However, to the extent that the significance of merely belonging to the community of motorists was relativized with the completion of mass motorization, the forms of expression of individual taste increased in importance in mass society.

Ever since the development of the specific form of the car, there had been low-volume brands and custom-made products that represented the desire for individualization of their wealthy owners. The distinctive features of such luxury models were expressed in an aestheticized form. At the same time, the tendency towards uniformity of serial mass production was accompanied by a desire for distinctiveness on the part of those in low-income occupational groups.

A selection of industrially manufactured extras, special designs and spoiler rings offered forms of expression for the individualization of serially-produced cars. With the help of parts offered by the manufacturer or the accessories industry, the owner was able to engage in a process of aesthetic appropriation and design. In the interior, the remodelling of the seats served not only to increase comfort or tailor them to the individual body, but also to furnish the car as an individualized space (cf. Csikszentmihalyi/Rochberg-Halton 1989: 47). In the 1950s, for example, it was common to decorate the dashboard with a vase for flower arrangements or with a talisman. The application of stickers with local emblems, advertising slogans or political statements developed into a symbolic form of ownership of the car. A hand-knitted cushion with the registration number of the vehicle on it, placed in the storage space under the rear window, was a common way of providing the mass product with a personalised, individual touch.

Symbolic Forms of Ownership

Power and Freedom

Driving practice is to a large extent determined by the ways in which the individual exercises control over his emotions. The driver's power over the process of acceleration, over a force that produces movement, has always been a temptation to act out non-rational, unconscious needs. Placing one's foot on the accelerator releases power that is many times stronger than the power of the driver's own body. The abreaction of pent-up aggressiveness has thereby become a far from rare experience in anonymous mass traffic, although the reasons for it are to be found outside the human-machine relationship, mostly in everyday relations. As a result, multi-layered forms of the symbolic demonstration of power over the machine figure among the forms of behavior in traffic: ostentatious acceleration and showy self-display on the part of the driver through a roar of the engine, driving at spectacular speeds and the intimidation of a weaker and slower vehicle by a stronger one are all expressions of the desire for power.

Expressions of the Desire for Power

There is no doubt that this applies in a gender-specific way predominantly to male road users.

On the other hand, because of its practical characteristics and uses, at a surprisingly early stage the car was associated with the symbol of freedom, which Bierbaum explicitly foregrounded back in 1906 in the following passage:

The railroad has turned the traveller into the passenger, the passenger who is passing through. But now we don't want to go past all the beautiful sights without stopping, simply because they are not envisaged in the timetable. We really want to travel once more as free men, choosing our destination freely, for pleasure purposes. [...] Travelling in a car involves not only a physical but also a mental massage, and this is precisely where its invigorating, freshening qualities lie. Passive travel is replaced by active travel. (Bierbaum, 1906: 333)

In the course of asserting this semantics of freedom, the slogan "unrestricted mobility for free citizens" (*freie Fahrt für freie Bürger*) became very popular in the 1960s. It assigned priority to the claims of individual expression over the restrictions imposed by the collective interests of society. Above all, the absence of a speed limit on the motorways was endowed with a pathos of freedom that included a radical lack of connection to the consequences of driving a car. In this situation, the emotional attachment to the car as object, the power over speed and the unlimited freedom to act it out were ideologically highly regarded. Whether this was especially true in Germany due to a lack of liberal political culture, as an expression of the achieved fulfillment of the stronger individual, would need to be explored further.

Semantics of Freedom

Power Over Speed as a Highly-Regarded Ideological Value

Concluding Remarks

Criticism of the automobile was kindled by the consequences of its quantitative expansion. In the 1980s, with mass motorization and mass tourism, the negative consequences, such as the increase in exhaust fumes, reached an extent (forest dieback, contribution to the heating of the earth's atmosphere, etc.) that provoked discussion about alternatives to the automobile, but at the same time revealed that only solutions that lie within the configuration of rapid mobility stand a realistic chance of acceptance and implementation. The structural requirement of mobility in a society based on the division of labor, but also the rationalization of individual mobility as a gain in enjoyment of life, are both so deeply rooted in modern culture that the need to restrict car traffic quickly reaches the limits of enforceability.

The electric car could contribute to rendering viable two approaches to restructuring the culture of individual mobility. This concerns, firstly, the idea of dividing urban space into zones authorized for gasoline-powered cars and zones where this is not the case. If the latter were to be approved for the operation of electric cars, this could significantly reduce resistance to the establishment of "prohibited zones". Furthermore, the electric car is suitable as a sensible component of integrated solutions, i.e. the combination of transport systems in the rail and road network. One would take the train in order to travel to the outskirts of the city, while individual mobility in the city centre would be dependent on the electric car, whose technical features make smooth and comfortable travel over short distances possible, without even minor hindrances. The transfer stations from the railway to the electric car could be comprehensively equipped with charging stations, so that the unavoidable downtimes of the cars in the park-and-ride system could be used for the battery charging process. In any case, new types of electric cars would have to be developed, geared to the practical values anchored in the configuration of modern everyday culture. A complete renunciation of the desire for distinction and the representation of status seems unlikely, given the mentality handed down in cultural history and the forms of contemporary culture in which the needs of car purchasers are inscribed and acted out.

Precisely because the automobile as an object of individual mobility is highly unlikely to be replaced, the electric car plays a special role in the transformation of older forms of mobility into culturally forward-looking forms of mobility, in order to contain the negative consequences of this "leading fossil" of modernity that have arisen in the object's more than century-long history (cf. Aicher 1984).

Object of Desire

3

The Electric Car in the Field of Political Power

Oliver Schwedes

Introductory Remarks

In the last three decades there have been various diagnoses of the times we live in, all of which attest to a profound social change. Since the general diagnosis of a "new complexity" by the social philosopher Jürgen Habermas (1985), various attempts have been made to conceptualize and encapsulate the new social conditions, starting with postmodernism or postfordism, through to the risk society, the thrill-seeking society and the network society. Even if each of these attempts at definition has captured an aspect of the current dynamics of development, none of the descriptions has prevailed in characterizing a new society as a whole.

In this chapter we adopt a different approach, while still considering the electric car in its socio-political context. We will focus on adequately describing processes of social development without examining them through the lens of a particular social paradigm. We can proceed from and build on the insight widely shared by the above-mentioned diagnoses, namely that we are still living in a capitalist society which, compared to all previous forms of society, is characterized by a particularly influential logic of economic development. As early as the middle of the nine-

---

The original version of this chapter was revised. A correction to this chapter is available at https://doi.org/10.1007/978-3-658-29760-2_8

O. Schwedes (✉)
Integrated Transport Planning, TU Berlin, Berlin, Germany
e-mail: oliver.schwedes@tu-berlin.de

© Springer Fachmedien Wiesbaden GmbH, part of Springer Nature 2021, corrected publication 2021
O. Schwedes, M. Keichel (eds.), *The Electric Car*,
https://doi.org/10.1007/978-3-658-29760-2_3

teenth century, Karl Marx and Friedrich Engels in the *Communist Manifesto* welcomed the great innovative dynamics of the capitalist processes of production as a progressive force, which demonstrates its effectiveness over and over again by revolutionizing social conditions: "All that is solid melts into air, all that is holy is profaned, and man is finally compelled to face with sober senses his real conditions of life and his mutual relations with his fellow men".[1] In similarly euphoric terms, at the beginning of the twentieth century the economist Joseph Schumpeter also praised this "process of creative destruction" as the essence of capitalist development: "Capitalism is thus by nature a form or method of economic change and not only never is but never can be stationary".[2]

The Electric Car in a Socio-political Context

The Nature of Capitalist Development

In contrast to this euphoric welcome given to the capitalist power of innovation, the economic historian Karl Polanyi had the negative social dimensions in mind when in the 1940s he diagnosed a compulsive process of destruction when looking back on nineteenth-century laissez-faire capitalism (Polanyi 1995). His explanation of the impoverishment of the proletariat associated with unrestrained industrialization was the far-reaching disembedding of the capitalist economic process from its social context. Polanyi described this unbridled economic dynamic as a "devil's mill", which crushes people if they are unable to keep up with economic development. He concluded that economic development must always be embedded in a cultural context that determines the direction and limits of the logic of economic development. The process of capitalist development is characterized by an interplay between the innovation-driven disembedding of the dynamics of economic development from traditional social contexts on the one hand, and their reintegration within the framework of newly-created social conditions on the other. Motivated by the "social question", the social reintegration of industrial workers finally took place thanks to the creation of social security systems in the welfare state. From then on, within the framework of public services, the state had the task of providing certain basic services which people in modern societies, unlike in the past, could no longer provide for themselves. In addition to various insurance services such as health, accident and pension insurance, these included in particu-

---

[1] Karl Marx and Friedrich Engels, *The Communist Manifesto*, ed. Jeffrey C. Isaac, Yale University Press, London and New Haven, 2012, p. 77.

[2] Joseph A. Schumpeter, *Capitalism, Socialism, and Democracy*, 3rd ed., Harper and Row, 1950, pp. 81.

lar infrastructure services such as connecting houses to the electricity grid, to the water supply and sewage disposal, but also adequate transport connections, which were to be provided by the public administration (cf. Schwedes/Gegner 2013).

The "Devil's Mill"

Mobility as a Basic Social Service

The analysis of capitalist development presented by Polanyi forms the starting point of the present contribution, which takes the electric car as an opportunity to examine the political parameters of a new *culture* of mobility. Accordingly, the global financial and economic crisis of 2008, as well as the crisis of the European Economic Union from 2010, form the provisional culmination of a phase of economic disembedding from social contexts, followed by a debate on the possibility of returning the (finance) economy to a newly-created social framework, with corresponding rules (cf. Peukert 2011). While this development involves ongoing major social upheavals, the ecological consequences have also been placed on the political agenda as a new challenge. On an international scale, the ecological limits of one-sided, economically-driven globalization are becoming increasingly apparent (cf. Altvater/Mahnkopf 2007). A special role in this is played by global transport flows, which are almost entirely dependent on oil as an energy source, the combustion of which is a major cause of climate change, with all its harmful consequences.

Ecological Limits of Economic Development

In view of this challenge, the electric car powered by renewable energy appears to be a promising vehicle for a shift from a fossil-based to a post-fossil mobility culture (Schindler et al. 2009). The present contribution examines the political prerequisites for the successful establishment of the electric car against the background of a necessary process of transformation in society as a whole. However, the overall trend cannot be fully predicted today; it cannot be extrapolated solely on the basis of economic laws, social structures, trends in technical development or cultural traditions. Rather, the thesis here is that the mobility culture of the future—including the electric car—must be decided politically. Against this background, it seems helpful to present an explanation of the possibilities and limits of politics in shaping the field of transport.

The Electric Car in the Process of Social Transformation

Who Is Talking About the Electric Car? What Is Being Said and Why?

Looking Back into the Future

In recent years, the media hype about the electric car has created the impression that this technical artefact is a tremendous innovation. The 125-year history of the electric car has rarely been the subject of discussion. Around 1900, 40% of cars in the USA were still steam-driven, 38% were electric and only 22% ran on gasoline. At that time, three technologies were still competing with each other and no decision had yet been made as to which would prevail. The engineer and historian of technology Gijs Mom (2004) shows in his socio-genesis of the electric car that the success of the internal combustion engine is due not only to its technological advantages but also to various cultural influences. According to his central thesis, the appeal of the internal combustion engine in comparison to the electric car lay precisely in its initial imperfections or susceptibility to faults. It was part of the adventure, in which men in particular were keen to demonstrate that they had mastered the machine themselves. In contrast, the—at the time—much more reliable electric car was stigmatized as a 'women's car'. Mom shows how the specific cultural context aided the internal combustion engine in its breakthrough. The gasoline car successively adopted the technical innovations successfully tested on the electric car, such as the closed chassis or the reinforced tyre casing. Ultimately, the invention of the electric starter helped to combine adventure with reliability and comfort in the internal combustion car. The "race and travel sedan" was born (cf. Knie 1997).

Media Hype: The Electric Car

If cultural conditions had been different, the opposite technological genesis would be conceivable today. If the positive image of the gasoline car as a race and travel sedan were to turn negative due to a cultural change in favor of ecological concerns, the automobile could undergo another metamorphosis and gradually develop from a solely gasoline-powered car to an electric car. The existing hybrid variants point in this direction. However, the transport sector today finds itself in a different situation than at the beginning of the twentieth century, when it was still open which path technological developments would take and which cultural characteristics would be decisive. Mom sees the established structures of the large-scale technical system of the race and travel sedan, with its multitude of participants and their specific interests, as a crucial additional hurdle in the development of the electric car (cf. Mom 2011). Unlike in the early days of automobile development, cultural change today has to be systematically accompanied by political decisions. In order to support the electric car as part of a sustainable transport development strategy, politicians would thus have the task of deliberately acting against the interests of the established players in the large-scale system of the gasoline-powered car and taking greater account of the new but so far marginalized players interested in the electric car (cf. Hoogma et al. 2002).

Metamorphosis of the Automobile

The Large-Scale Technical System of the Racing Sedan

E-mobility Hype in Comparison: The 1990s and Today

After the car with an internal combustion engine had established itself in the first half of the twentieth century, the electric car was repeatedly rediscovered. Nevertheless, the historical review reveals a genealogy of failure. Before the current rediscovery of the electric car, the last great e-mobility hype occurred in the 1990s (cf. Wallentowitz, Chap. 6). Despite the international attention, it disappeared as quickly as it had appeared. In light of this, two discourses on e-mobility promulgated by the media will be subjected to a comparative examination here in order to answer two questions in particular: first: are the chances of the electric car being successfully established greater today than 20 years ago? Secondly: do current trends indicate that the successful establishment of the electric car would contribute to the development of a sustainable transport strategy?[3]

A Genealogy of Failure

Common Features of the Two Discourses on Electric Mobility

At first glance, a comparison of the two discourses on electric mobility from the 1990s and 2000s reveals a number of similarities. In both cases, it was the media hype that came as a surprise, as an unexpected event for most of the participants.

---

[3] The following section is based on the study by Schwedes et al. (2011a).

Then as now, the discourse focused on the electric car. A closer look reveals that the electric car was never really 'dead', after existing alongside the car with an internal combustion engine until the 1920s and being replaced by the internal combustion engine in subsequent decades. Since the end of the Second World War, almost every decade there has been at least one study that confirms that the electric car has a great future. The news magazine, *Der Spiegel*, summed up the situation accordingly at the end of the 1990s: "The renewal of interest in electric cars is repeated in an almost cyclical sequence, with the same arguments for and against" (*Der Spiegel*, 26.04.1999). But only the discourse in the 1990s can justifiably be described as "hype", whereas previously there had merely been isolated news reports.

The First Hype About the Electric Car

Just like the current hype about electric mobility, the discourse of the 1990s was also determined by the coincidence of two important socio-political events. Firstly, the beginning of the 1990s was marked by an economic crisis which particularly affected the automotive industry (cf. Haipeter 2001). Secondly, this mood of economic crisis was associated with the climax of the ecological debate, which culminated in the first discussion on climate change at the beginning of the 1990s, characterized by the historian Joachim Radkau as the turning point of the ecological movement (cf. Radkau 2011: 488 ff.). As a result, the automotive industry in particular was plunged into a legitimacy crisis that had already been looming in the 1980s, with the news of forest dieback. Since then, the car has been identified as one of the principal environmental offenders and has increasingly come under scrutiny by the ecological movement, to the point where, at the beginning of the 1990s, the end of the automobile was proclaimed and demands were made for a fundamental change in the transport system (cf. Vester 1990; Berger/Servatius 1994; Canzler/Knie 1994).

Economic Crisis and the Ecology Movement

The automotive industry was on the defensive and engaged in discussions with its strongest critics. The conversion of the automotive industry from a car manufacturer to a mobility service provider was discussed. In this social atmosphere, the car was supposed to be—if not abolished—at least reinvented, with new drivetrain technologies in particular. At the same time, the US state of California adopted the Zero Emission Vehicle Program, which required every car manufacturer to produce at least 2% emission-free cars by the end of the 1990s, which at the time meant electric cars. Failure to comply would mean that companies in California would no longer be allowed to sell cars at all. This resulted in a further motivation for the German automotive industry not to remain closed-minded on the subject of electric cars any longer.

Legitimacy Crisis in the Automotive Industry

Nevertheless, the German automotive industry remained rather cautious, since it saw the electric car as a threat to its core competence, namely the internal combustion engine, which had been undergoing continuous development for 100 years. This hesitant attitude was criticized in the media. For decades, the only thing that had been done was to tinker lethargically and limit research to testing standard car bodies equipped with electric motors (cf. *Der Spiegel*, 08.07.1991). As early as 1991, transport experts saw no potential in electric cars to fully replace conventional cars; instead, they feared that the electric car could strengthen the trend towards second and third cars.

Politics and Electric Power Companies as Driving Forces

The driving force in the 1990s came from politicians, supported by the electric power companies. Both hoped to gain an advantage from the issue. While politicians were striving for an image gain through a symbolic politics of sustainability, the power companies wanted to open up a new sales market for themselves. The electric car was advertized as the perfect city car, with a limited range but enough to cover most routes in the city. The electric car was presented as the optimal niche vehicle for short distances, for commuting into cities, for health resorts and the like. It was also seen as an optimal alternative for fleet operators (cf. *Frankfurter Rundschau*, 06.05.1995, 11.05.1996).

In 1992, German politicians decided to further develop and test the electric car by supporting what was then the world's largest research project, on the island of Rügen, and individual federal states also participated in financing small pilot projects. However, as environmental discourse weakened and the automotive industry recovered from the economic crisis, interest declined in the bearer of hope for sustainable transport development. When the results of the Rügen project were made available in 1996 and showed that the electric car would not have a positive environmental effect due to the German electricity mix at the time and that its use could not be justified economically either, since the batteries were too expensive and also had only a limited range (cf. Voy 1996), the mood turned against the electric car.

Change of Mood Against the Electric Car

Politicians were no longer able to distinguish themselves as ecological pioneers with the electric car and the German automotive industry used the situation to say farewell to an unpopular alternative technology by blaming the failure on the battery manufacturers and their lack of effort on the development front. The energy industry, in turn, criticized the automotive industry for not getting involved in technological innovation for electric cars and for showing no willingness to adapt its conventional vehicles to the specific requirements of electric vehicles.

The current electric mobility hype began in 2007, under very similar conditions to those in the 1990s. This time, too, the beginning was marked by a global economic crisis that also affected the automotive industry, among others. With the debate on climate change, the economic crisis again coincided with a time of ecological crisis. Simultaneously with the onset of the financial and economic crisis, the *Intergovernmental Panel on Climate Change* published its fourth Assessment Report (IPCC 2007), which received worldwide attention. In this situation, the aim was on the one hand to support the automotive industry, an important economic sector for Germany, during the crisis. At the same time, German politicians found themselves compelled to avoid the impression that this was happening at the expense of the environment or the country's carbon footprint. As a result, in the two economic stimulus packages to promote the German economy, in addition to the so-called "eco-rebate", which everyone who exchanged their old car for a new one received, it was also decided to promote electric mobility with 500 million euros. Despite the relatively small sum (compared to 5 billion euros for the eco-rebate), the electric car nevertheless surprisingly quickly became a dominant theme in the media. As in the 1990s, the electric car was once again touted as a promising means of sustainable transport development. Politics was once again the driving force, assigning strategic importance to electric transport and, in the *National Electromobility Development Plan*, presenting it as a key factor in Germany's economic development, with the economic policy goal of achieving world market leadership in this area (cf. *Die Bundesregierung* 2009). The policy was actively supported by the energy industry, with the energy group RWE taking on a pioneering role with its high-profile activities. Again, the energy companies were hoping for new markets, while the automotive industry was adopting a rather passive, wait-

and-see stance (cf. Warnstorf-Berdelsmann 2012). Their basically skeptical attitude was masked by a strong media presence, which then, as now, was marked by
advertising campaigns. For years, the automotive companies had been announcing
that the first electric car ready for series production would soon be available and
that, as proof, the first prototypes would be presented at various automotive trade
fairs (Table 3.1).

The Second Hype Over the Electric Car

Hope for the Development of Sustainable Transport

These figures are belied by the small number of electric cars that are actually ready
for the market. All the initial pilot projects had to struggle with long lead times
before the first vehicles were on the road. Even at the beginning of the second wave
of funding in 2013, there were not enough electric vehicles available for the
planned research projects. As a rule, these were conventional vehicles converted to
electric vehicles at short notice. As in the 1990s, the automotive industry is primarily focusing on replacing the combustion engine with an electric drivetrain. Today,
as in the past, the special demands that the limitations of the electric car impose on
users remain largely disregarded. The high costs of batteries prevent private
purchases of electric cars in the short and medium term and the limited range is not
compatible with the favored strategy of replacing conventional vehicles with electric cars. The combustion engine cannot be replaced by the electric car, since the
latter is in fact a technical innovation the success of which depends on the willingness to use the automobile in new ways (cf. Ahrend and Stock, Chap. 5). However,
the perspective of the users is barely taken into account in publicly-funded projects.
Once again, the electric car has been hyped in the media as the savior, thus placing
it in the center of debates on transport policy, while the necessity of integrating it
into an overarching concept of transport policy remains unheeded.

Replacement of the Internal Combustion Engine by the Electric Drivetrain

The Electric Car as Savior

As was the case in the 1990s, the initially optimistic forecasts for the development
of electric transport are giving way to an increasingly sceptical view. Whereas the

Table 3.1 Key events

Year/period	Event	Brief description
1990–2001	Phase of very low oil prices.	Except for the price increases in the years 1990 and 1991, which were a consequence of the second Gulf War, the 1990s were marked by very low oil prices. Adjusted for inflation, the prices hovered around levels dating back to the beginning of the twentieth century.
1990	Economic crisis in the automobile industry.	The Fordist production system of the automobile industry entered a major crisis in the early 1990s, which also affected the industry's conception of itself. The automobile sector discussed a complete reorientation, away from automobile manufacturing to becoming a mobility service provider.
1990	Zero emissions vehicle (ZEV) program.	Reacting to high levels of air pollution, the California Air Resources Board (CARB) introduced the ZEV Program, which specified a quota for the production of zero-emission vehicles, stipulating that, as from 1998, 2% of all new vehicles sold annually had to be zero-emission vehicles. By 2003, the quota was supposed to increase to 10%. Given the state of the technology at the time, only battery-operated vehicles qualified as zero-emission vehicles.
1992	Launch of a large-scale test of electric vehicles on the island of Rügen.	Launch of a large-scale test of electric vehicles on the island of Rügen, sponsored by the then Federal Ministry for Research and Technology. At the time, the largest field trial worldwide, with 60 electric vehicles, most of them conventional automobiles that had been refitted with an electric drivetrain. The study focused on the everyday capability of the technology and its ecological effects.
1996	CARB cancels 1998 quota for ZEV.	CARB cancels the envisaged quota for the introduction of zero-emission vehicles in 1998. After discussions with representatives from the automobile industry, CARB agreed that that industry needed more time for the further development of the technology. The quota for the year 2003 remained unaltered.
1996	End of the large-scale testing of electric vehicles on the island of Rügen.	After the end of the large-scale testing of electric vehicles on the island of Rügen, the results were officially presented. Due to the ecological shortcomings, the results were negatively received by the Federal Ministry for Education, Science, Research and Technology and the Federal Environment Agency, as well as representatives from the press. The reasons for the negative result were the high energy consumption of the electric vehicles, as well as the unsustainable sources of electricity at the time.

continued

Table 3.1 continued

Year/period	Event	Brief description
1996	GM EV1 comes onto the market.	GM offers the electric car EV1 in selected regions of the USA as a leasing vehicle. In subsequent years other electric vehicles are released, such as the Ford Ranger EV and the Toyota RAV4 EV.
1997	Kyoto Protocol is adopted.	By signing the Kyoto Protocol the industrial nations commit themselves to reducing greenhouse gas emissions by 5% by 2008–2012 in comparison with the base years 1990 or 1995. The Kyoto Protocol meant that, for the first time, the industrialized nations were required under international law to meet binding targets for greenhouse gas emissions.
2000	Law giving precedence to renewable energy forms (Renewable Energy Act, EEG) comes into force.	The law is passed giving precedence to renewable energy in electric power generation, with the goal of decreasing reliance on fossil fuels, in the interest of climatic and environmental protection. The development of renewable energy is fostered through attractive remuneration of renewably-generated electricity and by obligating the network operators to purchase it. In subsequent years electricity generation from renewable energy forms gained considerably in importance.
2003	CARB weakens the standards for ZEV.	Following interventions by the automobile industry, CARB rescinds the plans for 10% emission-free vehicles by 2003. Instead of emission-free vehicles, vehicles with particularly low emissions have to be brought into operation.
2003	GM recalls EV1.	GM recalls leased EV1 and has them scrapped.
2006	Presentation of Tesla roadsters.	Tesla Motors presents the electric vehicle, the Tesla Roadster, after a 3-year development period. The small-scale production commences in 2008. The entirely battery-powered vehicle utilizes lithium-ion batteries for energy storage. It is the first larger application of this new kind of rechargeable battery in the automotive sector.
2007	Beginning of the financial and economic crisis.	Beginning of the US-American banking crisis, culminating in 2008 with the bankruptcy of the investment bank Lehman Brothers and triggering a worldwide financial and economic crisis. The automobile industry was especially hard hit. Due to its singular importance for the national economy, the German automobile industry was supported through various funding programmes (Stimulus Package I+II).

continued

Table 3.1 continued

Year/ period	Event	Brief description
2007	Intergovernmental panel on climate change (IPCC) Integrated Energy and Climate Programme (IEKP) adopted by the Federal government.	The publication of the 4th IPCC report attracts worldwide attention for the first time and also raises awareness on a national level for the issue of climate change. With the IEKP, the Federal government pursued energy and climate policy objectives. The measures addressed the domains of climate protection, the development of renewable energy and energy efficiency.
2008	Price of oil at an all-time high.	After substantial price rises since 2001, the price of oil reached an all-time high in July 2008.
2009	Setting of CO_2-emission standards for new automobiles.	The European Union sets regulatory limits for the average CO_2-emissions for new car fleets, to be introduced in a series of steps by 2015.
2009	Stimulus Package II with measures targeting electromobility.	Reacting to the financial and economic crisis, the Federal government enacts economic stimulus programs. The Stimulus Package II contains measures to promote research into electromobility, with funding of 500 million euros.
2009	National plan for the development of electromobility.	In the national plan for the development of electromobility, the Federal government set out the German research and development strategy in the field of electromobility. The goal was to promote research into and development of battery-operated electric vehicles in Germany, as well as preparing them for release onto the market. The contents and aims had been agreed upon by policy-makers together with participants from the economy and scientific research. The keypoints of the plan had already been discussed in 2008, in the framework of the national conference on strategies for electromobility.

Source: My own presentation

German government had previously expected one million electric cars on German roads by 2020, the *Institut der Deutschen Wirtschaft* now forecasts a maximum of 220,000 units (cf. IDW 2011).[4]

---

[4] It has to be mentioned here that the Federal Government's figures tacitly include hybrid vehicles, which comprise more than 50% of the total. The study by the Institute of the German Economy, on the other hand, only takes battery-powered electric vehicles into account. Nevertheless, this constitutes a clear change of direction.

Differences Between the Two Discourses on Electric Mobility

The manifold similarities between the two discourses should not tempt one to overlook the differences that also exist. In the current hype over electromobility, the climate debate is more important than at the beginning of the 1990s. This is reflected at EU level in the setting of ambitious climate protection targets. This also includes the introduction of regulatory limits for CO_2-emissions from passenger cars, which were reduced or delayed by interventions from the automotive industry (cf. Katzemich 2012), but can no longer be disputed in the long term. It is unclear when the development of the internal combustion engine will reach its limits and whether the electric car will be the only alternative for the future. In addition, the electric car is being discussed more intensely than in the 1990s in connection with renewable energies (cf. Billisch et al. 1994). This became possible after the introduction in 2000 of the Renewable Energy Sources Act (EEG), which gave priority to renewable energies and led to the rapid development of renewable energy sources (cf. Scheer 2010). Thus the argument concerning the unsustainable energy mix, which was still influential at the time, has now been significantly weakened. Although it is still rightly pointed out that the electric car hardly emits less CO_2-emissions than a comparable car with a combustion engine (cf. Öko-Institut 2011), the prospect of a fleet of electric cars powered by renewable energies is nonetheless tangible today. At least in Germany, this trend in favor of renewable energies was probably encouraged by the change in energy policy initiated as a result of the nuclear catastrophe at Fukushima and could thus also strengthen the prospects of the electric car.

The Electric Car and Renewable Energy

Environmental Track Record

Another major difference is the steady rise in oil prices over the past few years. Although the oil price has not yet led to a change in transport behavior, it has revived the debate about the finite nature of fossil fuels. Under the heading "peak oil", there has been increasing discussion in recent years as to how long global oil resources will last, on which the transport sector is more than 90% dependent. This question arises all the more urgently when considering the rapid mobilization of the emerging countries (cf. Schwedes/Rammler 2012). Against this background, the electric car powered by renewable energy is increasingly being discussed as a contribution to independence from increasingly scarce oil reserves. This has given rise to a completely new argument, which was not part of the discourse on electric mobility in the 1990s, and which today may be the decisive argument in favor of the electric car. As a result, energy policy goals are becoming more important, along with environmental ones.

Independence from Oil

An additional argument in favor of developing the electric car is the successful establishment of hybrid technology by Toyota in the 1990s. It is increasingly regarded as a transitional technology and makes a gradual electrification of automobile transport seem possible. Not least of all, the successful market development of hybrid technology has contributed to a rethink in the automotive industry, which is beginning to enter into new alliances with players outside the industry, such as the battery industry. The development of the lithium-ion battery constitutes a technological advance compared to the 1990s, although it does not constitute a breakthrough in battery technology. However, the leaps in development of the technology have encouraged the renewed emergence of the hype (cf. also Linzbach et al. 2009: 16 f.). As a result, players that previously had no connection to the sector, such as the energy industry, but also certain suppliers to the automotive industry, have gained in importance. In contrast to the 1990s, a constellation of actors is emerging that could possibly lead to a change in the balance of power in the transport sector. Even though it is still the same protagonists who are determining the direction of the discourse, their importance within the discourse formation has changed.

Changes in the Balance of Power

The current discourse shows a much more complex and tighter network of relationships than in the 1990s. This reflects in particular a change in the political parameters. The Renewable Energy Sources Act (EEG) resulted in a rapid expansion of renewable energy. This development has been given an additional boost by the intended phasing out of nuclear energy. In this context, the electric car takes on a whole new meaning. It is no longer just the second-best alternative to the internal combustion engine, but an integral component of a new post-fossil energy concept in the current discourse. In this energy policy strategy, the energy industry in particular is becoming increasingly important.

New Players

The so-called "new players" are also benefiting from this change in the importance of the electric car. In most cases, they come from the automotive supply industry, which also defines their formal status. Others, such as battery manufacturers, will become part of the supplier industry. It is well known that small and medium-sized supplier companies have been suffering for a long time from their one-sided dependence on large automotive groups. So far they have not appeared as actors, because they acted less than they reacted. To the extent that the electric car gains in importance, the one-sided dependence is relativized in favor of the new players. A high level of competence in the field of electrical engineering plays a much more important role in electric cars and strengthens the corresponding suppliers, such as Bosch, by changing the value chain—in some cases so much so that these companies are starting to develop their own electric cars. To the extent that new players with innovative sales concepts in particular are gaining in importance, such as *German Rail* (*Deutsche Bahn*) or *Better Place*, companies that have a genuine interest in the establishment and further development of the electric car, one can expect an attenuation of the pronounced wait-and-see attentism of the established automotive industry.

As a result, with the development of electric transport the constellation of players has expanded and the balance of power between the players has changed (cf. Fig. 3.1).

The almost unlimited power of the automotive industry, which was still able to exert direct influence on politics in the 1990s, was relativized by the growing importance of the energy industry and the new players. At this point, however, it has to be expressly emphasized that this assessment of the situation is restricted to the discourse on electromobility, which is only one strand among many in the debate on alternative drivetrains. The ostensible loss of influence by the automotive industry in the field of electromobility must itself therefore be viewed as an interim assessment, since it relates to a set of issues whose future development is not yet clear.

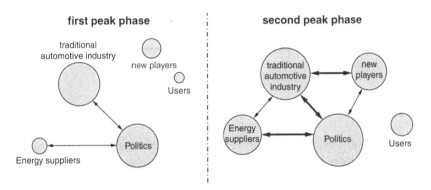

Fig. 3.1 Importance of (size of the circle) and relationship between (thickness of the arrow) the actors. Initial boom phase; Second boom phase; Traditional automobile industry; New actors; Users; Politics; Energy provider (Source: own presentation)

Shift in the Balance of Power

An additional incentive for the further development of electric cars is the growing pressure of international competition. The forecasts for the development of electric cars in China, the feared competitor in recent years, have now been revised downwards significantly (cf. WI 2012). There are other countries, such as the USA or France, that are promoting development and also supporting it with financial incentives. Against this backdrop, the situation in Germany could also change soon, should a buyer's premium be introduced here as well, once the first German series models are on the market.

Lastly, the research projects today cover a broader spectrum and far more extensive funds are available now than in the 1990s. Even if one gets the impression that the initial euphoria over electromobility is fading, there is no indication of an overall standstill. In contrast to the 1990s, the topic will be pursued in the coming years (cf. GGEMO 2011).

Broad Research Agenda

Classification of the Current Discourse on Electric Mobility

After identifying similarities and differences between the two discourses on electromobility, I will now undertake a classification of the current hype on electromo-

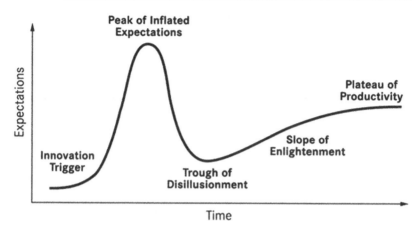

Fig. 3.2   The technological hype-cycle (Source: Fenn/Raskino (2008: 9))

bility. The aim is to answer the two key questions raised at the outset: (1) will the current discourse on e-mobility, in contrast to the developments in the 1990s, prove to be durable this time around? And (2): if so, what contribution would this make to the development of sustainable transport?

In classifying the current hype on electromobility, I adopt as a guide the so-called hype cycle (cf. Fig. 3.2), according to which hype ideally goes through five phases: the first phase is initiated by a technological trigger (innovation trigger). An event, such as the start of a research project, generates increased public interest in a particular technology. In the case of electric transport, it is the public sector, which promoted pilot projects both in the 1990s and again since 2009, that contributed to increased attention to electric transport. The second phase is then characterized by exaggerated expectations (peak of inflated expectations). Against the background of the successful testing of the new technology, enthusiastic forecasts predominate, while their "teething problems" are largely downplayed. Media attention reaches its high point. Both the electric mobility hype of the 1990s and the current hype went through this phase.

Then comes the third phase of the technology hype, namely the "trough of disillusionment". With the increasingly realistic assessment of the new technology, taking into account the existing shortcomings, disillusionment sets in, which is associated with decreasing media coverage. Looking back at the electric mobility hype of the 1990s, it can be seen that it remained stuck in the "trough of disillusionment", so to speak. In light of this, the current hype raises the question of whether it will reach the fourth phase (the "slope of enlightenment"), which is characterized by a realistic assessment of the potentials of the technology, in which the advantages and disadvantages are soberly weighed against each other. As a result, the new technology again attracts growing media attention. With regard to the current electromobililty hype, it is not yet possible to say with any certainty whether this fourth phase has already been reached and whether the "trough of disillusionment" has been crossed. At the very least, both the fundamental willingness of politicians to press ahead with this issue and the genuine development of electric cars for their intrinsic value by the car industry (cf. Keichel, Chap. 4) indicate that the phase involving more objective examination of the new technology will not end this time by being prematurely cut short. Lastly, the fifth phase marks a stabilization of public interest in the new technology ("plateau of productivity"), with the degree of attention being determined by whether the technology establishes itself in the mass market or remains a niche product. The current electric mobility hype repeatedly gives the impression that the establishment of electric cars has already been decided. From today's perspective, however, there are many indications that we are close to the peak. The first signs of disillusionment are emerging (the targets for 2020 will clearly not be reached, very few vehicles are on the market, sales figures are low, further problems such as charging infrastructure are becoming apparent). Some observers are predicting the "trough of disillusionment", while others offer a more appropriate assessment of the situation. In any case, it has not yet been decided whether the purely battery-driven electric drivetrain will prevail as an alternative drive type.

In the Trough of Disillusionment

It's a Tie!

Discussion of the Results

A central finding of the study is that the topic of electromobility did not fade in the 1990s due to technical failures or under-development, as is repeatedly claimed. From a transport policy point of view, this is an important insight. For instead of reassuring oneself with the simple reasoning that in the past the electric car failed due to under-developed battery technology, it is necessary to explain why electric transport did not establish itself in the 1990s, even though electric cars had been developed that possessed comparable performance parameters and were already marketed as perfect city cars at the time.[5]

Technological Determinism

Assessing the fortunes of the electric car in the 1990s by fixating on the technology stems from a fundamentally narrow view of the subject. Then, as now, the electric car was primarily viewed from the point of view of technical feasibility, where the performance characteristics of the conventional car serve as the yardstick. There is a danger of overlooking other factors that have a bearing on the success or failure of technological innovations. For example, the argument that the German automotive industry in the 1990s criticized the battery industry for not having fully-developed batteries does not seem convincing from today's point of view. Even back then, electric cars were already capable of ranges that would have been sufficient for the vast majority of journeys in urban agglomerations. The response from the battery industry at the time, namely that the automotive industry showed no interest in a successful development of the electric car, since it was limiting itself to redesigning conventional vehicles instead of designing corresponding bodies for the electric car, is a valid point that is again being discussed in the current debate. Ultimately, the Federal Environment Agency rejected the electric car in the 1990s, pointing out that it offered no ecological advantages due to the high proportion of fossil fuels in the German electricity mix. This argument was as convincing back then as it is today, because today it is still emphasized that the electric car only contributes to a positive environmental outcome when the electricity comes from

---

[5] The historian of technology Gijs Mom goes so far as to explain the failure in the 1990s by resorting to conspiracy theory (cf. Mom 2011). The conspiracy theorists repeatedly refer to the film "Who Killed the Electric Car?" (an excerpt can be found here: https://www.youtube.com/watch?v=r75lqbA0uMM accessed 22.09.2019). However, the film does not reveal a conspiracy, but rather identifies a number of groups of players who, from their specific point of view, spoke out against the electric car for good reasons. These include the majority of the US population, who were not willing or able to dispense with the four-door, gasoline-fuelled sedan, in view of the social conditions at the time. Thus, no conspiracy was needed to thwart the electric car, but rather the social conditions at the time were sufficient, where individual groups of players, such as the oil and automotive industries, with their respective particular interests, benefited from the gasoline-fuelled car. In reality, the electric car failed in the 1990s both in the USA and in Germany due to a lack of political transparency, which prevented an enlightened public discourse that would have made it possible to objectively weigh up the advantages and disadvantages of the electric car within the framework of a sustainable transport development strategy instead of being dominated by powerful special interests.

renewable energy. Even back then, politicians could have come to the conclusion that the electric car is able to make a positive contribution to sustainable transport development under certain political conditions. These energy and transport policy framework conditions should have been defined by politicians. But instead of combining the development of the electric car with a new energy policy in favor of renewable energy, the politicians, together with the automobile industry, decided against the electric car.

Battery Industry Versus Automotive Industry

Energy and Transport Policy Framework

A new situation did not arise until the EEG (Renewable Energy Act) was adopted in 2000. This political reorientation led to a rapid expansion of renewable energies and undoubtedly contributed to the fact that the electric car can now be envisaged as a contribution to a sustainable energy strategy.[6] Instead of criticizing the energy companies for relying on the inefficient electric car, as they did in the 1990s, politicians have created a framework to ensure that the large corporations no longer block the expansion of renewable energies, as they did in the 1990s (cf. Becker 2010).[7] As a result, the balance of power between the players involved in the discourse on electric mobility has shifted. This provides a clearly positive answer to the question of whether we are experiencing a qualitatively new discourse on electric mobility compared to the 1990s.

A New Quality in the Current Discourse on Electromobility

In the current phase, the electric car is primarily an industrial and energy policy issue. The central arguments are the preservation or expansion of the international

---

[6] This is an example of how politics makes a difference. Contrary to the popular assertion that differences between the political parties have been levelled out, energy policy shows that the coalition between the Social Democrats and the Greens at the end of the 1990s initiated a turnaround that the coalition between the Christian Democrats (CDU-CSU) and the Liberals (FDP) was neither willing nor able to achieve. The CDU/CSU-FDP coalition's initial attempt to reverse the nuclear phase-out at the expense of the expansion of renewable energies made this clear once again.

[7] Under the CDU/CSU-FDP Federal Government, however, the transformation of the energy system was again increasingly contested (cf. Kemfert 2013).

competitiveness of the German automotive industry as well as the scarcity of oil resources and, consequently, the necessity to become independent of fossil fuels in the long term. By comparison, transport and environmental policy arguments play a lesser role. On the contrary, the strong focus on technological innovation in electric cars has led to sustainable transport policy measures increasingly being overlooked, measures that could already be implemented in the short and medium term and that are relatively inexpensive compared to the cost of electric cars (cf. Petersen 2011). This means that the strong orientation towards technology comes at the expense of an impact orientation, such as the sensible concentration on reducing CO_2-emissions. In the current discourse on electric mobility, the electric car serves as a metaphor for sustainable urban transport development, but behind it lies the primary economic policy goal of global technical market leadership. The many negative environmental effects that would also not be solved by electric cars, especially in urban agglomerations, have hardly been addressed so far (cf. Schwedes et al. 2011b). The fact that sustainable transport development requires different concepts of transport and different transport behavior, which has to be directed at using the car—including the electric car—less, plays at best a marginal role in the current discourse.

The Electric Car as a Metaphor for Sustainable Urban Transport

After the uproar about the financial crisis and climate change, which originally justified the promotion of the electric car, concerns about dependence on dwindling oil resources became a strong new argument on the agenda. Since then, the electric car as a technological innovation has served as the bearer of hope for sustainable development, whereas electric mobility as a component of a comprehensive sustainable transport policy, with the goal of creating a new mobility culture, is hardly discussed any more.

A Comprehensive Sustainable Transport Policy

With regard to the second guiding question of the present chapter, namely to what extent the current discourse on electromobility supports sustainable transport development, it must therefore be pointed out that the electric car is currently understood primarily as a technological innovation and not as a component of an integrated, comprehensive transport policy aimed at sustainable transport development.

Conclusions

The starting point of this investigation was the diagnosis of a process of profound social transformation. While the challenge at the beginning of the twentieth century was to reconcile unbridled laissez-faire capitalism with social issues, at the beginning of the twenty-first century the ecological question, a global issue, entered the political agenda. The appropriate handling of the ecological question is, however, just as politically controversial as social issues were in the past. In the political field of transport, the resulting challenge is seen in the transition from a fossil-based to a post-fossil mobility culture. The electric car could play a central role, provided it is powered by electricity from renewable sources.

Far-Reaching Process of Social Transformation

Against this background, we examined the relevance of the current hype concerning electromobility, motivated by the observation that there had already been a similar hype in the 1990s, which disappeared from the political agenda just as quickly as it had appeared and remained largely without consequences. In light of this experience, we endeavored to clarify to what extent the current discourse on electromobility is characterized by a new quality, due to possible changes in social conditions. Irrespective of whether the current hype over electromobility may prove to be more consequential than its predecessor in the 1990s, the second question focused on its how sustainable it is. In contrast to the current debate, where the electric car is mostly presented unquestioningly as a contribution to sustainable transport development, we examined the current discourse on electric mobility from the point of view of the extent to which it contributes to supporting such a development.

The Electric Car : A Contribution to Sustainable Transport Development?

The comparison of the two discourses initially showed astonishing similarities. In both cases, the trigger was an economic crisis that affected an important industry in Germany, namely the automotive industry. In addition, the economic crisis was linked to contemporary environmental discourse. In both cases, the electric car was the one point on which all participants were able to agree, with more or less commitment. Although the automotive industry saw the electric car as a competing technology and was accordingly sceptical, it was able to re-work its profile during the crisis with a partial commitment to the technology and also received financial support. The electricity companies, on the other hand, were understandably positive about the electric car, since they associated its spread with a new sales market. Politicians, for their part, recognized the electric car as an opportunity to distinguish themselves in two ways: with the financial support of the automotive industry, they were able to make a contribution to overcoming the economic crisis and at the same time position themselves in the environmental discourse thanks to a potentially environmentally friendly technology.

The Electric Car as a Compromise Formula

The fact that politicians took a rather opportunistic stance in the 1990s became clear at the moment when the electric car came under increasing criticism, particularly from the automotive industry, due to its—on balance—negative environmental impact, resulting from the electricity mix at the time. Instead of taking the quite justified criticism as an occasion for a new energy policy, politicians joined forces with the automotive industry and stepped back from promoting the electric car. The situation now is quite different. Now that politicians have embarked on a new energy policy path and renewable energies have gained a significant portion of the electricity mix, the electric car appears in a different light. Within the framework of a development strategy directed towards a post-fossil mobility culture, the electric car could play an important role in the long term. This opens up a potentially new horizon for the electric car that did not exist in the 1990s, due to the lack of political support at the time.

A Potentially New Horizon

The comparison of the two discourses on electromobility results in a central insight into transport policy. The new quality of the current discourse on electric mobility is not the result of economic or technological considerations, but rather the result of a political decision in favor of renewable energy. There are two things about this realization that matter: *First*: even in the current discourse, neither the repeatedly cited economic limitations (expensive batteries) nor technological shortcomings (insufficient battery capacity) is decisive for the successful establishment of the electric car. Rather, it can be assumed that a political decision, however it is justified, will be the pivotal factor. *Secondly*: the development of the electric car is currently not driven by transport policy considerations, but is rather discussed within the framework of considerations regarding energy and industrial policy. This partly answers the question of the extent to which the current discourse provides indications for the sustainable development of electric transport.

Energy and Industrial Policy Take Precedence Over Transport Policy

It is remarkable that energy and industrial policy arguments are achieving in the shortest possible time what has been called for in environmental policy for decades. As has been shown, the energy and industrial policy arguments become convincing against the background of an impending shortage of fossil fuels, in particular crude oil. The focus is not on negative environmental effects such as climate change, which has been discussed periodically since the 1970s, but has remained a relatively weak argument to this day. The strong argument, on the other hand, is the threat to the economic system posed by dwindling oil supplies. Transport, which is more than 90% dependent on oil worldwide, plays a central role here.

Technology-Oriented Strategy for Finding a Solution

In view of the motivation provided by energy and industrial policy, it is not surprising that the discourse on electromobility is not driven by environmental considerations either. Rather, the electric car is the expression of a technology-oriented strategy to find a solution. Especially in the automotive sector, the traditional idea is that social problems created by the technical artefact of the automobile can be solved by technological innovations. All these examples are part of a popular strategy of efficiency that focuses on solving problems through innovative further development of proven technology (cf. Weizsäcker et al. 2010).

Replace Old Resource Dependency with a New One

Of course, the efficiency strategy is only one of three strategies that originally characterized the path to sustainable development (cf. Huber 1995). The second is the consistency strategy, which is aimed at reusing or recycling materials (cf. Braungart/ McDonough 2011). The aim is to develop chemical substances that people use which, like natural substances, can be repeatedly fed into material cycles without ever being discarded as waste. In contrast to the efficiency strategy, the consistency strategy has so far hardly played a role in the development of electric cars. Rather, it is still completely unclear how the batteries can be disposed of or recycled. From the point of view of sustainability, this is problematic in two respects: on the one hand, it concerns the handling of toxic substances and, on the other hand, scarce resources such as rare earths (cf. Blume et al. 2011). If the electric car is not simply going to help replace old negative environmental effects with new ones and replace old resource dependencies with new ones, but is rather going to be part of a sustainable transport development strategy, then the consistency strategy has to be promoted.

Goal: Behavioural Change

Finally, a sustainable approach includes the so-called sufficiency strategy, which aims to change people's behaviour (cf. Princen 2005; Stengel 2011). It is based on the insight that, while the efficiency and consistency strategies can contribute to a more efficient and effective use of natural resources, the ever-increasing consumption is counteracting the savings. This has been particularly evident in the transport sector, where, for example, the effect of saving fuel thanks to more economical engines has repeatedly been counteracted by the growing volume of traffic (Banister 2008: 20 f.). The sufficiency strategy therefore aims to influence people's transport behavior in such a way as to reduce the overall volume of traffic. In the case of the electric car, the aim would be not only to replace the internal combustion engine with an electric motor, which would ideally fulfil the efficiency and consistency strategy, but also to use the electric car differently and less than cars with an internal combustion engine. This requires comprehensive transport policy measures aimed at changing transport behaviour.

From Energy to Transport

As the development of the electric car to date has shown, its contribution to sustainable transport development depends on appropriate political decisions. As a result of the political decision for the development of renewable energies, the electric car today makes a contribution both in the sense of the efficiency as well as the consistency strategy, since electric motors have a much higher degree of efficiency and, moreover, are powered by renewable energies. However, in order for the electric car to contribute to sustainable transport development, the necessary political decisions in favor of the sufficiency strategy are still lacking. It would be the task of politicians to make the electric car the starting point of a transport turnaround, analogous to the energy turnaround. In other words, the successful development of electric transport in the interests of sustainable transport development will in future require a political programme aimed at a new mobility culture.

Completely New Possibilities

On the Design of the Electric Car

Marcus Keichel

Introductory Remarks

If one enters the term "electro-mobility" together with the sentence fragment "completely new possibilities" into a search engine, one gets an astonishing number of hits. With one click numerous quotations appear, expressing the spirit of optimism surrounding the electric car. Spokespersons from energy companies extol the sales opportunities for electricity, suppliers of automobile parts see business potential in the field of battery development and engineers rave about the options in vehicle construction. Although the subject is by no means new, electro-mobility now seems to have been positively received on a broad scale, now that the political will to implement it is at hand.

Automobile Designers

This also applies to automobile designers. In articles and interviews there is no lack of emphasis on "completely new possibilities" when it comes to their views on the

M. Keichel (✉)
Läufer+Keichel, Berlin, Germany
e-mail: marcus.keichel@laeuferkeichel.de

electric car.[1] The reasons for this open-mindedness clearly have to do with the specific reality of car designers' work. Automotive design is a multi-layered profession, at times fraught with contradictions and conflicting goals. First of all, the designers are required to have a high degree of creativity and to be skilled in articulating their designs. Firstly because the professionals and customers generally expect a high degree of formal sophistication and visual surprises from a new car. Secondly, the majority of car owners want to signal their commitment to progress and modernity by buying a new car. For this purpose, the designers must provide a new model with design elements that represent the aesthetic spirit of the time and the state of technical development. A keen sense of contemporary formal codes is just as important as comprehensive knowledge of the latest technical components, production processes and materials.[2] And thirdly, designers must base their designs on the brand and product policy of the company for which they work. This has to do above all with the competitive situation on the automotive market: in order to be competitive, a car manufacturer must produce large quantities, i.e. for a wide range of customers. This can only succeed if the brand profile—i.e. the well-balanced interplay of product design and quality, pricing, marketing and service—corresponds with the tastes, lifestyles and self-images of as large a group of car buyers as possible. This is the reason for the diversification and partial convergence since the 1990s of historically rather distinctive brand profiles. Since then, car brands traditionally identified as "conservative" have been striving to create "sporty" or "youthful" images for their products and vice versa. Manufacturers of mid-range cars add luxury cars to their range, while premium brands offer "entry-level models". Such strategic brand differentiations must be reflected by the designers in their product concepts. As a rule, they do this by first inscribing a distinctive basic symbolism into the manufacturer's various hierarchically structured model series. In the case of a manufacturer with a sporty image, the body design of both the lowest and top model series in the hierarchy is given a sporty touch (vertical product family).[3] For each of these model series, the designers develop a large selection of accessories and design applications (Economy-Line, Sport-Line, Executive-Line,

---

[1] For example, Lutz Fügener in an interview entitled "New Possibilities". The automotive designer is convinced that "electric cars [sooner or later] will have their own unique design language", http://www.berlinonline.de/themen/auto-und-motor/autotechnik/1003157-61213-kühlervoneautosschaffenplatzfürideen.de.html, accessed: 19.05.2012.

[2] The latest headlamp technology traditionally plays a role here. Cf. "Des einer Freud, des anderen Light", in: *Auto Motor und Sport*, No. 13, 2012, p. 124 ff.

[3] At Volkswagen, for example, the Polo, Golf, Passat and Phaeton model series constitute a vertical product family; at BMW the 1-series, 3-series, 5-series, 6-series and 7-series, etc.

etc.), most of which are combined in equipment and accessory packages, with which the various models in such a series can be produced (horizontal product family). A portfolio differentiated in this way creates a variance of product motifs that enables manufacturers to reach buyers who do not belong to the core clientele.

Creativity and Skill in Articulating Designs

Convergence of Brand Profiles

Automobile as Bearer of Symbolic Messages

The criteria for modern automobile design indicated here suggest that the special demands for creativity placed on car designers have to do primarily with the traditional significance of the automobile as a bearer of symbolic messages. The car has long been—and now more so than ever—a "means of expressing ideas",[4] which essentially revolve around the car owners' need for distinction and self-stylisation. As creators of symbolic expression, automobile designers orient themselves comparatively strongly to the habitus of the creative artist.[5] When beginning to develop the expressive basis of a design, car designers, like figurative sculptors, generate partially abstract images and forms which can nevertheless be charged with meaning. Technical or functional restrictions are largely ignored, because questions of practicability or utility values still play a subordinate role in these phases.[6]

Habitus of the Creative Artist

Restrictions

However, in the course of product development, restrictions increasingly come to the fore until they finally assume a dimension that is greater than in most other fields of industrial design: strict safety regulations, high rationalisation pressure

---

[4] A definition of symbol, originally by Clifford Geertz (1987), here quoted from Ruppert, Chap. 2.

[5] The term "artist's habitus" is fundamental: Ruppert 1998.

[6] The comparatively strong orientation of car designers towards the artist's habitus is expressed, among other things, in the fact that they work in the medium of expressive drawings or volume models, especially at the beginning of product development. In these media, the character of the object ranks before the character of the product; what is conveyed is an "automotive" expression rather than depicting an already completed car. This expression is only transferred to the technical artefact "car" in the product development process.

(platform concepts and the use of identical parts), aerodynamic criteria, etc. constitute a strong force field of constraints within which the designers have to implement their aesthetic-symbolic concept.

The tension that stems from an artist designer having to adapt appreciably to the hard criteria of the automotive industry explains the basic openness of car designers to the electric car. The potential for creative leeway opens up because the alternative drive concept promises no fewer but partly different restrictions. More than before, aesthetically and conceptually innovative design solutions for automobiles appear in the realm of what is feasible, innovative forms of use and adoption of automobiles are conceivable. Overstating the situation somewhat, one could say that under the current conditions, with politicians and industry having decided to promote and develop purpose-designed electric cars (cf. Wallentowitz, Chap. 6), designers have the rare opportunity to become a driving force in the development of a new and sustainable mobility culture.[7]

Potential Scope for Design

Designers as Initiators in the Development of a Sustainable Mobility Culture

Interest in Knowledge and Methodology

In the following, I would like to explore the question of whether and in what way automobile designers use this opportunity. I deliberately want to concentrate on the design of the vehicles themselves. One can only agree with authors who maintain that, if the project is to be ecologically successful, the design of e-mobility must include the development of an intelligent infrastructure that ultimately reduces private car ownership and individual traffic. There is no doubt that the simple existence of electric vehicles, possibly as an addition to the gasoline-powered car fleet (i.e., as a second or third car), does not solve the ecological problems of individual transport in any way. Less comprehensible, however, is the assertion by the same

---

[7]Representatives of the automobile industry postulate—albeit in abstract form—a connection between the development of purpose-designed electric cars and an imminent change in mobility culture. For example, BMW CEO Norbert Reithofer is quoted in a broadsheet newspaper article on the status of electric car developments: "Tomorrow's mobility will be different from today's", *Süddeutsche Zeitung* of 12–13 January 2013, p. V2/11.

authors that the design of the infrastructure must take priority over the design of the cars themselves.[8]

Design of the Vehicles

Historical experience shows that, when it comes to the automobile, we must assume a particularly strong, in part mythically charged man-machine relationship (cf. Ruppert, Chap. 2). Whether a new automobile is accepted or not depends to a large extent on the aesthetic-symbolic content of its design. The long-term tradition of this special relationship makes it seem more than improbable that a more dispassionate perception of the electric car, liberated from symbolic ballast as it were, could be produced by concentrating the creative energy of the designers on the "system" of electric transport and not on its main objects, i.e. the cars. On the contrary: the electric car is also a car, and it is reasonable to assume that, beyond a completely new symbolic potential (it is perfectly conceivable that the electric car could be transformed into *the* central emblem of the post-oil era), it will continue to be acquired and used as a medium for conveying conventional messages (individuality, distinction, dynamism, modernity, etc.).

Electric Car as an Emblem of the Post-oil Era

The mere necessity of having to create acceptance for a new kind of vehicle, which is not in every respect superior in its practical usefulness, would seem to make it imperative to take the design of the product seriously. But this is not the only reason. If the forecasts are correct, then in a few decades' time we will have to deal with a large number of electric cars in our living environments, especially in urban areas. Their aesthetic appearance will have a formative influence on the atmosphere in cities. This means that the design of the vehicles is important even for those people who do not have a driver's licence; for they, too, are entitled to a product symbolism that is conducive to the public ambience.[9] The values on which such a

---

[8] Stefan Rammler, Director of the Institute for Transportation Design (ITD) at the University of Fine Arts Braunschweig, in an interview in *Design Report*, issue 3/2012, p. 31.

[9] The architecture critic Niklas Maak points to the connection between car design and urban culture. In his perception, the design of modern SUVs is calculated to symbolically transform public space into "battle zones". Niklas Maak: "Die heisse und die kalte Stadt", in: TU München und Bayrische Akademie der Schönen Künste (ed.): *Die Tradition von morgen. Architektur in München seit 1980*, Munich 2012, p. 29.

symbolism is based can and indeed must be debated. However, it is the designers' own responsibility to create these values—nobody can help them.[10] Conversely, it is imperative that transportation designers do not go it alone in developing innovative and user-friendly infrastructures, including pay-as-you-go systems, battery service stations and smartphone apps, but instead collaborate with creative actors from other professions such as transport planners, architects, software developers, etc.

The i3 Project from BMW

As a further focus of my presentation, I would like to take a closer look at a single development project, using it as a case study. I have in mind the i3 project of Bayrische Motorenwerke AG. Concentrating on this example, which in the best case can be considered representative and thus meaningful beyond this individual instance, is intended to provide a certain depth of focus. This is necessary when it comes to understanding the significance of approaches to design and the contexts in which they originate. Image interpretations based on free association, which are not uncommon in journalism on design, are inadequate when exploring these areas and involve the danger of imprecision and misinterpretations.

The BMW i3 project seems to me to be suitable as a case study for a number of reasons. First of all, because it is based on a differentiated concept that makes it seem distinctly possible that this electric car can be developed into a leading product of e-mobility. This is because it is a compact electric car that is designed for operation over short and medium distances.[11] Given the state of development of

---

[10] Unfortunately, the marginalization of the aesthetic-symbolic dimension of creative work in the German discourse on design remains a stubbornly entrenched cliché. In the *design report* (Issue 3/2012: Electro-Mobility), for example, there is an illustration that is accompanied by a particularly unfortunate comment by the editors: "First the concept, then the form: sketches for a small van with electric drive" (ibid. p. 32). Apart from the fact that there can be no concept without form, this cliché persistently ignores the reality of the dynamics of creativity, within which ideas can emerge in an intuitive play with forms and images, the content of which goes beyond the purely pictorial.

[11] The charging of the BMW i3 as an icon of e-mobility was indicated at an early stage in the form of its media representation. For example, in a large-format article in the *Süddeutsche Zeitung* on the status of the "Electric Offensive" (12–13 January 2013). The article is illustrated with a total of four illustrations, whereby the BMW i3 is shown several times larger than the electric cars of other manufacturers. the stylistically comparable i8 model was also euphorically commented on in the design report as early as 2012: "Innovation in the fast lane: the BMW i8 Spyder as an elaborately designed image carrier" (issue 3/2012: Elektromobilität. p. 38).

batteries, this is a realistic application scenario (cf. Wallentowitz, Chap. 6). But in addition—and this is probably more important—the BMW strategists take the aesthetic dimension of the electric car seriously. The pre-release illustrations made it clear that the designers wanted to appeal to the emotions of their customers, who are enthusiastic about technology and cars. The BMW i3 differs visually from gasoline-powered cars in a way that will probably be positively received by customers. Thanks to the visual presentation of the materials and technologies, as well as the partial abolition of the classic vehicle structure, the car looks different from those currently on the roads. At the same time it is not *strikingly* different and in a comprehensive—i.e. also symbolic—sense, it remains a car. Lastly, the project is also suitable for closer examination because it is well documented. Due to its advanced stage of development, information is available on both the development objectives and the end product itself.

The Aesthetic Dimension of the Electric Car

Without wishing to pre-empt the analysis, it should already be noted that the BMW project is strongly influenced by classic patterns of thought and imagination in the automotive industry. The players are naturally committed above all to the economic success of the products and are thus strongly involved in the discourses customary in the industry about current trends, the activities of the competition, international developments, and so on. The scope for reflection on the longer-term cultural significance of design strategies is naturally limited in this practice—at least there is very little that is made public. In the concluding part of this chapter, I would like to sound out some possibly useful reflections, going beyond the dynamics of the day-to-day business. Of particular interest here is the question of what designers can learn from the history of their discipline, i.e. to what extent sustainable design concepts from the past offer stimulating potential for the development of successful and culturally valuable mobility products in the future.

The BMW i3 Electric Car: A Case Study

The design of a new car is decisive for its success. The extent to which this is the case can be judged by the extremely strict practice of secrecy. Until just before the official presentation at a trade fair or a similar event, manufacturers usually take great care to ensure that no one learns what the new models look like.[12] A modern design, in keeping with—or even ahead of—the *zeitgeist*, creates a time advantage over competitors and the later they can react to this advantage, the longer it lasts.

This basically simple merchant's wisdom seems to have been suspended in the case of electric cars. In any case, BMW already published the first comprehensive photo documentation on the two projects i3 and i8—the latter being a two-seater sports car with hybrid drive—in 2011, i.e. two or three years before the planned market launch (cf. *Der Spiegel*, 10.09.2012). As astonishing as this communication policy (which deviates diametrically from current practice) may seem, the reasons for it are obvious: the manufacturer assumes that it cannot expect the same a priori acceptance for electric models as for conventional cars. For this reason, with the pre-release of the new models, the company is pursuing early marketing with the aim of eliminating reservations concerning the still unfamiliar product at an early stage.[13] BMW marketing relies heavily on the visual attractiveness of the product design. As with all of its products, the company sees this as the key to its success: for its entry into e-mobility, it is focusing on a clientele with a need for symbolic differentiation. In this sense, corporate strategists assume that the design of electric cars must differ from that of conventional models. However, in order not to unsettle the regular clientele and their image of the BMW brand and its products, the company decided to create the sub-brand BMW i. The electric models are now sold under this brand name.

Visual Attractiveness of the Product Design

The importance that the company attaches to the design of electric cars can also be seen in the fact that in media communication about BMW i projects, it is less the company management or even the sales department that has a say than the head of design. Adrian van Hooydonk is responsible for explaining the new i-products in interviews, but above all the design criteria on which they are based. These inter-

---

[12] In the test phases of technical development, test drives in public spaces are always carried out using externally heavily disguised prototypes (so-called "Erlkönigen").

[13] In the words of automobile designer Lutz Fügeners: "After all, […] [the customers] have been used to conventional cars for many decades and need some time to get used to [electric cars]", http://www.berlinonline.de/themen/auto-und-motor/autotechnik/1003157-61213-kühlervoneautosschaffenplatzfürideende.html-und-motor/autotechnik/1003157-61213-kühlervoneautosschaffenplatzfürideende.html Access: 16.10.2019.

views are a highly valuable source when attempting to understand the BMW strategists' approach to the subject of electric cars.

The Design Guidelines

In June 2011 Adrian van Hooydonck was interviewed by the automotive trade magazine *Auto Motor and Sport*.[14] The editors asked the designer about general developments in the field of electric transport, but in the course of the interview they quickly focused their interest on the design features that distinguish BMW i electric cars from BMW's gasoline-powered cars:

> **AMS**: Are the cars really different? **van Hooydonck**: Yes, we are even creating a new brand for it. The streetscape will change drastically in two years due to the i3 megacity vehicle and the i8 sports car. The cars look ultra-modern, and you have the feeling that the future has arrived.
> **AMS**: What's so different about them?
> **van Hooydonck**: The construction, the use of carbon, the lightness and the aerodynamics, which the car also expresses visually. There are clear differences to the BMW brand, which traditionally consists in expressing sporty elegance.

In this brief introductory passage to the interview, van Hooydonck names the central terms on which the design of BMW electric cars is primarily based: *modernity*, *aerodynamics* and *lightweight construction*. This has to be considered surprising, since they simply repeat the long-established, guiding principles of conventional automobile construction. Examining the terms in turn: *modernity* is perhaps one of the "original characteristics" of the automobile par excellence, and even doubly so. On the one hand, the car as a technical artefact was already integrated into the overall process of modern civilization when it was invented at the end of the 19th century (cf. Ruppert, Chap. 2). As a technically highly complex, industrially manufactured machine that soon far exceeded human performance capabilities, the car quickly became a leading product within this process and has remained so to this day. Indeed, the reputation of an industrial nation is measured to a significant degree by the state of development of its automotive industry. On the other hand, since the 1920s at the latest, the automobile has been *the* symbolic object of a modern bourgeois lifestyle, within which the experience of individual mobility, freedom (of movement), the dissolution of spatial boundaries, dynamism and acceleration has developed into a fundamental value (cf. Leggewie 2011). Ownership

---

[14] Unless otherwise stated, the following quotes are taken from this interview: http://www.auto-motor-und-sport.de/news/adrian-van-hooydonck-im-interview-der-bmw-designchef-ueber-die-e-zukunft-3815578.html

of a modern car has been a symbol of participation in modern life for almost one hundred years now.

Modernity, Aerodynamics and Lightweight Construction: Guiding Principles of Conventional Automotive Engineering

On the topic of *lightweight construction*: at the very beginning of automobile construction, when engines still had comparatively little power, lightweight construction also played a similarly important role, as one would expect with the construction of horse-drawn carriages. Cruising speed and range depended decisively on the weight of the carriage. So it was no coincidence that the first automobiles followed the constructional principles of carriage construction (cf. Ruppert, Chap. 2). Although the importance of lightweight designs initially declined with the development of more powerful engines, they experienced a renaissance following the first oil crisis at the beginning of the 1970s, in connection with the aim of reducing fuel consumption.[15] In the field of sporty automobiles, the development of lightweight vehicles has been an abiding concern. Their enduring presence in the car landscape made the terms *lightweight construction* and *high performance* into synonyms of automotive engineering. Lastly, *aerodynamics* is also a long-established design criterion for automobiles. Since the 1960s, the body shape even of ordinary automobiles is the result of systematic research into the reduction of air resistance.[16] The initial focus was on increasing the top speed, before other effects such as the reduction of fuel consumption and wind noise became more important a little later.

How is it that the head of design at BMW, when asked what is so special about the design of electric cars, first and foremost lists well-known principles from the field of conventional automobile construction? There is some evidence that designers and product strategists are consciously or unconsciously resorting to proven design criteria out of concern for the acceptance of electric models. *Modernity*, *lightweight construction* and *aerodynamics* are positively connoted terms for car lovers, and it is obviously this target group that the company wants to reach. The way in which this is to be accomplished is especially vividly illustrated by the design criterion of *aerodynamics*. It is noteworthy here that it is not only not a new design criterion, but also an ambivalent one in connection with the BMW i3, since this electric car is primarily intended for the city, i.e. for traveling at lower speeds:

---

[15] The VW Golf, introduced in 1974, weighed 810 kg and had a DIN consumption of 5.2 l/100 km at 90 km/h (model variant Formula E).

[16] The best known example of this is probably the NSU RO 80, introduced in 1967 and designed by Hans Luthe (cf. Aicher 1984: 41).

AMS: The i3 is a small city van? How do you get good aerodynamics?

van Hooydonk: The BMW i3 will not be a van, but a new, modern vehicle for urban mobility. A sedan has inherently better aerodynamic characteristics than a car with a one-box design. But we were also concerned about spatial efficiency and the possibility of offering as much vehicle space as possible in a small area assigned to traffic. Beyond that, we succeeded in achieving good aerodynamics.

Although the editors of the trade journal do not ask any further questions about the importance of an aerodynamic body for a city car (mega-city-vehicle), van Hooydonk indirectly reveals the answer shortly afterwards: "Aerodynamics will basically become visible," he explains. With this remarkable formulation, it becomes clear that it is less the real, effective advantages of low drag (such as a longer range within a battery charging cycle) that are at the forefront, but rather *symbolic* concerns. The designers are primarily interested in the image of an aerodynamic vehicle. Thus they continue to highlight speed, dynamism and increased mobility as attractively fashionable—but in reality still conventional—"values" of car culture and thereby perpetuate aesthetic models that comprise part of the fascination of the car for their clientele.[17]

Highlighting Speed, Dynamism and Increased Mobility

The same applies to the issue of *lightweight construction:* the designers also want to "make the reduced weight of the vehicle visible in the design" (van Hooydonk): Since the design measures for this, such as the use of carbon-fibre reinforced plastics in the chassis area, for example, can hardly be visually highlighted, the designers are turning to alternative measures: the roof structure is to be provided with light-coloured materials and large glass surfaces. If one disregards the fact that the glass looks light but is actually a material with a high specific weight, the question nevertheless arises: why highlight it? Weight reduction has advantages in terms of range and performance, but is it therefore a value that has to be symbolically communicated, especially using such expedients? As with aerodynamics, this seems to

---

[17] The earliest example of this would be the "streamlined form", symbolically charged as early as the 1930s, which quickly became an aesthetic cipher for the fascinating intoxication of speed. For some time it had a style-forming character in automobile design, and this although it had in fact produced no noteworthy advantages—the "Volkswagen" developed by Ferdinand Porsche around 1930 (the later KdF car or VW Beetle) is certainly the most prominent example of an automobile designed in the spirit of the streamline.

be a heightened but essentially conventional symbolic measure, where "lightweight construction" is traditionally synonymous with "high performance".

To preclude misunderstandings: *modernity, aerodynamics* and *lightweight construction* are, of course, legitimate design criteria for electric cars. Electric cars are also cars, and it is obvious that a considerable number of the lessons learned from the development history of the gasoline-powered automobile are relevant for the design of electric cars. Our interest, however, is in the new design of electric cars, i.e. the "completely new possibilities" mentioned earlier. They don't say so explicitly, but the motoring journalists of *Car Motor and Sport* seem to have noticed that the BMW head of design does not reveal much about this in the first part of the interview. In any case, they ask some follow-up questions:

> **AMS:** Is the appearance of cars with electric drive changing fundamentally?
>
> **van Hooydonk:** With the i3 the driver is practically sitting on the battery and the electric motor. This also changes the proportions. [...]
>
> **AMS:** Can you reveal any more of the tricks for expressing "zero emission" visually? **Van Hooydonk:** Compared to a BMW Z4, an electric car looks more serene. We're talking about clean mobility here, and this can also be expressed by a clean design of the surfaces.

Here we learn something about the specific design approaches for the BMW i models. Van Hooydonk suggests that something novel is to be created both in the area of the vehicle body construction and on the vehicle surface. As far as the body of the vehicle is concerned, this involves the changed design conditions brought about by the electric drive: the motor and battery of the i3 form a geometrically flat unit that can be placed underneath the passenger compartment.[18] Such a structure would certainly not make sense with gasoline engines due to their height and the temperatures they develop. We do not know whether this body structure leads to real advantages in vehicle use. However, it can be assumed that, compared to a conventional car of the same overall length, there is more cargo space available, since the area in front of the passenger compartment is no longer needed for the drive system. This could turn the (former) engine compartment into additional cargo space.[19]

---

[18] The statement that "the driver is virtually sitting on the battery and the electric motor" is misleading, since according to the published illustrations the motor is positioned above the rear axle.

[19] On the other hand, it can be assumed that the i3 will be higher than a comparable gasoline-powered car due to the superimposed arrangement of drive, energy storage and the passenger compartment. This could mean that a larger number of such cars, especially when parked in confined urban spaces, could obstruct the lines of sight and appear bulky. A similarly regrettable phenomenon can be observed in connection with the current proliferation of luxury SUVs in inner-city locations.

With their question about the specific symbolic messages that the design of electric cars is intended to convey, it is interesting to note that the editors of *Car Motor and Sport* simply assume that the central issue is meant to be the emission-free nature of electric drives. Van Hooydonk also seems to confirm this by declaring that he has opted for a "quiet" and "clean" design of the surfaces as a sign of "clean mobility". Apart from the statement that only small air inlets had been provided at the front of the vehicle in order to symbolically express a drive that requires no cooling, the topic does not become more concrete in the further course of the interview.

Specific Symbolic Messages of the Electric Car

In summary, Adrian van Hooydonk outlines the design of the BMW electric cars in such a way that they transmit the fundamental principles of modernity, aerodynamics and lightweight construction. Furthermore, changes in vehicle proportions (higher and shorter) occur as a result of the construction. As a symbolic representation of "clean mobility", the designers, in van Hooydonk's words, focused on a serene appearance ("clean design") and small air inlets. Lastly, they find it imperative to ensure the brand reference by stylistic means: to this end, they work with the "sharp lines" typical of BMW and the application of the so-called BMW "kidney"—originally a two-part radiator opening—as a design feature with a long tradition and high recognition value.

One would not be doing Van Hooydonk and the product managers an injustice by pointing out that they have been very cautious in drawing up the design criteria catalogue for their electric cars: they have avoided innovative features that would have been conceivable and would have served to clearly demarcate the vehicles from conventional automotive culture.

The Design

After evaluating a source text that provided the guiding principles of design and statements regarding concrete design measures, I would now like to turn my attention to the visual sources, namely the pre-release illustrations and photographs of the BMW i3. The main interest lies in the similarities with, and divergences from, Hooydonk's remarks, as well as those design features that remain unmentioned in the interview.[20]

---

[20] The images published pre-release are computer representations and photos of prototypes that had been presented at road shows since 2011.

Overall Appearance: Modern and Dynamic

As one might have guessed, when one looks at the designs, one finds what one expects as well as surprises. First the expected: a first glance at the illustrations makes it clear that the BMW i3 is a compact car whose overall appearance is modern and dynamic by today's standards. The announced "visible" aerodynamics are easy to distinguish: the low-slung body, the strongly-inclined front end with windscreen, the steeply vertical rear end with the "tear-off" edge. Glass is indeed generously and consistently employed. Not just the roof is glazed, but also the upper two-thirds of the doors are made of glass, in an extension of the side windows. Thus the "waistline" of the car appears to be clearly shifted downwards and the passenger compartment is a correspondingly open space. Generally speaking, the body of the car is much more transparent than is the case with conventional small cars. The use of glass as a design element is so far-reaching that the classic vehicle structure seems to have been abolished: the B and C columns are located behind the surrounding side glazing and thus recede into the background, visually speaking—the vehicle roof appears to be floating. The intensive use of state-of-the-art materials and technologies in body parts, headlights, wheel rims, etc. contribute to the radically modern appearance of the i3. Van Hooydonk had alluded to all this, and it is correct (Fig. 4.1).

Visual Noise

And now to the unexpected: First of all, the large number of curved vehicle edges and the graphic complexity of the many dividing lines around the doors, lights,

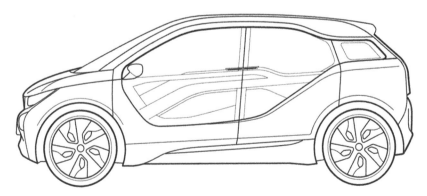

Fig. 4.1 "Clean design" The BMW i3 in side view (Source: Keichel)

cooling openings, etc. are bewildering. These characteristics are difficult to reconcile with the statement that the i3 is visually "serene" or "clean". Rather the opposite is the case: apart from the small hood, there are no body surfaces that are not divided. Almost everywhere internal forms bulge out or hard dividing lines cut the surface areas into countless subzones. The multitude of abstract gestures creates a visual noise, in which a main creative motif is difficult to discern. The hard two-tone body (basic colour light silver, hood black) reinforces the impression of a fractured form.

Next, a number of design features stand out that van Hooydonk doesn't even mention in the interview, even though they are supposed to identify the i3 as an electric car: I'm referring to a number of bright blue coloured decorative applications, which are obviously supposed to symbolically express the "clean drive". These include the blue-painted BMW "kidney" at the front of the vehicle, a blue decorative ring around the BMW emblem, ring-shaped blue inserts in the tyre flanks, a blue border for the charging port and a pronounced blue decorative strip on the side below the vehicle doors. It is not clear why van Hooydonk does not mention these highly conspicuous applications. Possibly because they are too unspecific as symbols, in any case other manufacturers also use blue applications and shimmering blue light elements to visualize electricity as a "clean form of energy".[21]

"Face" of the Car

Finally, however, the biggest surprise: the front end of the BMW i3. It is the 'face' of the car, like no other design element it determines its expression, and it is—naturally—the result of intensive design work. How does an electric car that is conceived as a trend-setter actually look out into the world? This question may well have occupied the designers for a long time. Unfortunately we don't know anything about the controversies regarding the physiognomy of the BMW i3, because van Hooydonk doesn't say a word about this either. But one can see the result, which is as clear as it is succinct: grim! With its eyes half closed in the form of the headlights cut off at the top, ears stuck on, expressed by the corners of the front end drawn strongly towards the rear and the large, wide-open nostrils (the BMW "kid-

---

[21] The front of the Audi E-Tron, for example, lights up blue when the battery is being charged, http://www.berlinonline.de/themen/auto-und-motor/autotechnik/1003157-61213-kühlervoneautosschaffenplatzfürideende.html, accessed: 19.05.2012.

ney"), the designers have created a "face" that seems frozen in a threatening gesture.

In combination with the many taut parts of the bodywork, in which one can perceive a muscular physicality without much effort of the imagination, the car seems to convey something like a permanent readiness to fight—as if the designers had been interested in depicting an attack dog. Amazing as it may seem, the primary symbolism of this electric car is an aggressive one. All the other messages—modernity, lightweight construction, aerodynamics, etc.—are conveyed in the context of this symbolism and ultimately contribute to its intensification. How does this symbolism come about? "Clean mobility" on the one hand, declaration of war on the other—where is the connection? (Fig. 4.2).

Aggressive Symbolism

There is some evidence that the designers and product strategists of BMW i also played it safe in this central point of vehicle design out of concern for the acceptance of their new product and therefore chose a conventional path. Because the "grim look" is another familiar pattern from the car world, but in this case not so much a historical as a contemporary one. In January 2012, architecture and design critic Niklas Maak described his impressions of his visit to the NAIAS International American Motor Show in Detroit:

> The front ends of the vast majority of contemporary cars seem to depict hysterization. They look like the masks of a Greek tragedy; one sees fear-distorted, panic-stricken grimaces, wide open radiator mouths, headlights in the form of glowing frown lines,

Fig. 4.2 The front of the BMW i3 (Source: Keichel)

latticed metal jaws, as if the vehicle was fed not on petrol but on hooved animals, swallowed whole […]. (Frankfurter Allgemeine Zeitung, 12.01.2012: 29)

The thesis that the overall expression of the BMW i3 design is a less original or even trend-setting design achievement, but is rather oriented towards current trends, gains in plausibility if one compares recent design studies from BMW. A retro design study from 2008 is particularly striking. Back then, the company presented the BMW M1 Homage Car. This is an interpretation of the BMW M1 two-seater sports car from 1978 that is strongly influenced by the spirit of the times, i.e. playful, but above all bursting with power (Fig. 4.3).

During its production period until 1981, the original M1 was not a great sales success. As the first uncompromising BMW road sports car of the post-war era, however, it has developed over the years into a cult "vintage" car, for which collectors now pay top prices of 150,000 euros and more. The fact that it was present as a racing car in the relevant media of the late seventies and early eighties and was driven by prominent racing drivers such as Niki Lauda may have contributed to the myths surrounding the M1. The current retro-design study is a marketing-motivated attempt to link up with the mythical M1 and highlight the image of the brand as a manufacturer of sporty automobiles. The design of the homage car relies on the tension between the comparatively understated wedge shape of the historical model, which is still alluded to, and the numerous dynamically designed inner forms and styling edges ("lines of force") that now surround it. Of course the tires and rims are bigger and wider, the gaze of the headlights more "poisonous" and the air intakes more numerous and bigger than the original. A number of nostalgic quotations can also be discerned, such as the gill-like air intakes on the hood, a common style element in the sports car design of the seventies.

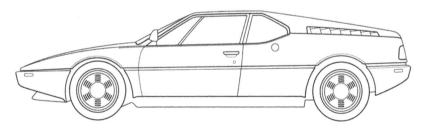

Fig. 4.3 The BMW M1 (Design: Giogietto Giugiaro, 1978) (Source: Keichel)

Retro-Design Study

Nostalgic Design Concept

The BMW M1 homage car is interesting because it gives the impression that this rather nostalgic design concept was the inspiration for the future-oriented electric car projects i3 and i8. The similarity, especially with regard to the expression of the front end, is unmistakable, and where the visual relationship to the i8, which, like the M1, is designed as a two-seater sports car, is particularly clear (Fig. 4.4).[22]

This finding makes it clear that design concepts—even those for resolutely new products—do not fall out of the wide blue sky. Rather, they are situated in the continuity of longer-term developments and are created by people who think and work in certain contexts. The market-ready BMW i3 and i8 electric cars are products of an automotive group in which car designers develop sporty automobiles with great success. Most of them are stylistically "mannered", but above all they are endowed with an expression of strong self-assertion, to put it mildly. This means—it should once again be emphasized—that they do not constitute a minority trend in the current aesthetic zeitgeist of automobile design, but rather belong to the mainstream: perhaps a little more so than cars from other manufacturers, a modern BMW is, symbolically speaking, an ego-bloating product.[23] And that is exactly what the BMW designers and product strategists want the new electric cars to be.

Expression of Strong Self-assertion

Ego-Bloating Product

Critique
How to evaluate such a finding? The chance of "completely new possibilities" wasted on the first attempt?

---

[22] The commentary accompanying one of the photos with which the BMW M1 homage car was presented on *Spiegel Online* in April 2008 explicitly referred to its aggressive appearance: "Düsterer Kumpan: bei dieser Beleuchtung wirkt das BMW M1 Hommage Car beinahe furchteinflößend". ["Grim sidekick: with this lighting equipment the BMW M1 homage car looks almost fearsome."]. http://www.spiegel.de/fotostrecke/bmw-m1-hommage-car-keil-fuer-die-zukunft-fotostrecke-31014-6.html, accessed: October 2, 2012.

[23] This striking formulation, concerning the connection between aesthetic experience and feelings, comes from the conductor Rupert Huber. In a radio interview he described Richard Wagner's compositions as music that releases "egoblähende Kräfte" (ego-bloating forces) and provokes in the listeners an emotional "detachment" from their fellow human beings. *Deutschlandfunk*, April 2008.

Fig. 4.4 BMW M1 Homage
Car (top) and BMW i8
(bottom) (Source: Keichel)

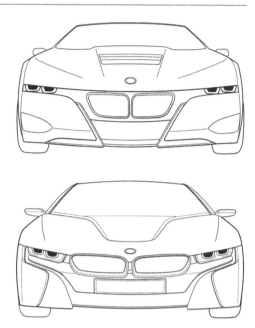

Correlation: Aggressive Car Design and Aggressive Behavior in Traffic

If one considers the problematic side of the BMW i3's product symbolism, the first thing that comes to mind is a possible connection between aggressive car design and aggressive behavior in traffic. Niklas Maak at least presumes an aggressive message, which the drivers send into public space via such a design:

> The car greets other road users with its maw-like radiator, reinforced with luminous, chromed teeth, which [...] signals that the driver regards public space as a place in which eating and being eaten are on the agenda. (Frankfurter Allgemeine Zeitung, 12.01.2012: 29)

The extent to which the combination of brand and product symbolism actually affects the social behavior of motorists needs to be studied. However, where empirical data are available, they seem to confirm a connection. The ADAC reported in its membership magazine of September 2012 that 93% of all motorists had already been victims of aggressive behavior in traffic on several occasions (ADAC Motorwelt 2012: 21). The member survey showed the interesting result in this context that a good 50% of the respondents are of the opinion that the most aggressive drivers are sitting behind the wheel of a BMW. Significantly fewer respondents had the impression that the most aggressive behavior came from Mercedes (32.3%) or Audi drivers (25.9%).[24] If one assumes that the performance characteristics of these brands of car do not differ dramatically and that a comparable number of them are present on the roads, the survey result supports the assumption that there is a connection between brand and product symbolism on the one hand and social behavior in traffic on the other. On the basis of the apparent trend alone it would be worth considering alternative brand and product design concepts—especially with regard to the electric car. However, the main problem associated with the product symbolism of the BMW i3 is a different one.

Collective Fixation of Modern Societies on the Car

The negative consequences of the strong collective fixation of modern societies on the automobile as *the* object of mass mobility have been known for a long time and they can be itemized as follows: environmental pollution, land consumption, traffic jams, loss of urban quality of life and impairment of health due to lack of exercise, etc. Measured against ecological, cultural and health standards, it would make sense to combine the introduction of electric transport with the aim of reducing individual car traffic. In Germany, a quarter of all trips under one kilometer and half of all trips under five kilometers are made by car, although traffic lights and a shortage of parking space largely negate the time saved, compared with bicycles (Leggewie 2011: 96). For a voluntary renunciation of short trips by car to become a reality—a renunciation which in the form of a renaissance of cycling can certainly be experienced as a physical, sensual and emotional gain—it would be of decisive importance to lessen the mental fixation of a large part of the population on the automobile. In this context, the design concept of the BMW i3 does the exact opposite: it displays the car as a fetish object. On the one hand, because its symbolism upholds the potential function of the automobile as an "ego prosthesis;"[25] on the other hand, because it fits seamlessly into a prominent social dynamic in which social advancement on the one hand and, on the other hand, symbolic demarcation from those lower on the social scale are

---

[24] Multiple answers were possible.

[25] This term was coined by Wolfgang Sachs, in: *Die Liebe zum Automobil. Ein Rückblick in die Geschichte unserer Wünsche*. Reinbeck near Hamburg, 1990.

prized aims and common practice. The design of the BMW i3 is calculated to offer a financially well-off middle class—the car currently costs around forty thousand euros in the basic version—the option of a gain in prestige, in keeping with the motto: some of us can afford the extravagant electric car, others cannot. Experience has shown that many people find such a "purchased" gain in prestige attractive. At the same time, as a vehicle, the i3 is at its most effective in situations that offer the driver an audience, namely on short trips in urban spaces. There is much to be said for the prediction that an electric car like the BMW i3 will reignite the demand for a kind of mobility that is very likely the most dispensable of all. If this were to happen, it would be contrary to the ecological objectives of the Electric Mobility Initiative, because the entire life cycle of an electric car, from manufacture through operation to disposal, cannot be described as emission-free—especially if the electricity for the operation of the vehicle does not come from renewable energy sources.

The Car as a Fetish Object

Social Advancement and Symbolic Demarcation

Fuelling the Need for Mobility: Irreconcilable with Ecological Objectives

It would of course be too one-sided to focus exclusively on the problematic side of the BMW approach to design. The professionalism of the work is just as undeniable as the refinement displayed by the numerous innovative solutions for design details, for example in the area of the tail lights. And it is undoubtedly positive that a car manufacturer like BMW takes people seriously as aesthetic beings, thus acknowledging the cultural and symbolic dimension of the automobile. BMW has to sell its products. The company is therefore dependent on communicating with its customers and engaging with their complex needs. On this point, a private enterprise seems to have an advantage over independent research institutions. This is because mobility solutions that are technically highly innovative and offer real practical advantages often come from such institutions, but major doubts remain as to their practical enforceability because the researcher-developers take insufficient account of the cultural context of the automobile. The significance of the car as a symbol is just as seldom taken into account as the individual emotional investment in the driving experience.[26] Those who do not take these "soft" factors seriously will not be successful with their concepts—even if they are the right ones in terms of practical use. The latter qualities in particular are not fully realised in the BMW i3, for the reasons outlined above, but it still has a real chance on the market as a "car". Should it become a success, there will be no denying its significance as a pioneer project in creating acceptance for the vehicle. That much, at least, needs to be said for it.

Cultural Context of the Automobile

Conclusions

The current electro-mobility initiative is not a unique instance of environmental, transport or energy policy regulation. There have been similar initiatives in the past. Past experience has shown that "hype" surrounding the electric car can soon lead to disillusionment and, as a result, to the loss of genuine interest. This was most recently the case at the beginning of the 1990s (cf. Schwedes, Chap. 3 and Wallentowitz, Chap. 6). But history does not necessarily repeat itself and there is reason to believe that things will turn out differently this time, even though the euphoric tone in the reports from 2010 to mid-2011 has meanwhile given way to more sober descriptions of the situation.[27]

A New Form of Taking Pleasure in Cars

A measure of the success of the current initiative would include consumers not just accepting the advent of electric cars but really wanting them. The BMW projects i3

---

[26] One example of such an approach is the "EO smart connecting car" project, which was developed at the Bremen site of the German Research Center for Artificial Intelligence (DFKI). The concept is based on the daring assumption that the drivers of electric cars are willing to give up their individual driving experience in order to link their vehicle with similar vehicles of other drivers to so-called "road trains". The advantage of this solution would be lower energy consumption (shifted to the individual vehicle) and a greater range.

[27] The *Süddeutsche Zeitung* reports on the former SAP manager Shai Agassi, for example. In an almost full-page interview on March 15, 2010, the newspaper presented him as a business pioneer who had recently succeeded in acquiring $700 million in investment capital for his e-mobility company, Better Place. On October 13, 2012, the newspaper reported under the heading "Battery Empty" that Agassi had to resign from his executive position at Better Place because the company had fallen far short of the sales forecasts for electric cars in Israel and Denmark.

and i8 could be a contribution on this level. But "success" also means that the electric car will help society move forward in terms of mobility. The criteria used to measure supra-individual progress in mobility culture are today partly different than in the past. In part, this even involves correcting the undesirable developments connected with mass motorisation since the 1960s. A core task consists in loosening the mental fixation of a large part of the (mainly male) population on the automobile as a kind of mobility fetish. The aim is not to make a case for a lifestyle that is anti-car or even anti-pleasure. On the contrary: with the introduction of the electric car, it could be possible to provide a large group of people with a new form of pleasure in cars and driving. The prerequisite for this would be encouraging and facilitating a less "driven" relationship to the product, free of excessive symbolic ballast. Only those practices that are free of compulsion and redundancy can ultimately be experienced as truly joyful. The much-quoted motto "less is more" applies in this instance: driving an electric car will be especially enjoyable if it does not atrophy into daily routine. For example, if you regularly commute to work and don't do so by car, the emotional experience of a car journey is something special for you. The development of public transport services into an attractive alternative would of course be a necessary, but not sufficient, criterion for such a change in mobility culture. Equally crucial would be a change in the collective perception of the car as an object. What can designers contribute to this change? What does a car look like that promotes change in the sense of a more relaxed human-object relationship?

Change in Mobility Culture

In 2011, the political scientist Claus Leggewie presented his much-acclaimed polemic "Mut statt Wut" ("Courage instead of Anger"), in which he encourages us citizens to actively participate in shaping social change. It also contains a chapter in which the sustainability criteria for mobility are discussed. Leggewie lists four goals: improving the means of transport, altering the flow of traffic, reassigning individual transport and forgoing mobility (Leggewie 2011: 98). The author argues in favor of smaller, lighter, slower, less powerful vehicles, better adapted to real transport needs (ibid.: 102). He thus outlines a canon of measurable design criteria that consciously sets itself apart from the current practice in conventional automobile construction and that could be easily reconciled with the systemic features of electric cars. Of course, it is more difficult to delineate the soft factors, for example aesthetic features. Niklas Maak made an attempt, even going so far as to invoke the aesthetic appeal of a past icon of car design:

It is time for a new DS—which was so aesthetically ground-breaking because back then it already had everything that hybrid and electric cars would need today: it was streamlined, had disguised fenders, an elongated body that had been designed in a wind tunnel, creating the impression of floating silently in the air, instead of rumbling over the ground accompanied by explosively loud engine noise. (Frankfurter Allgemeine Zeitung, 12.01.2012: 29)

Smooth and Silent Movement Through Space

The Citroen DS still impresses today with its streamlined shape and it is certainly true that it expresses something that fits the character of electric cars: an almost frictionless and therefore quiet movement through space. Apart from that, however, little seems to speak for the "goddess" as a leitmotif for the design of electric cars. When it was first released in the mid-1950s, the Citroen DS was a very large, extremely elaborately built and not exactly inexpensive prestige product, a stately car in two senses: on the one hand the President's official car and on the other—like the Eiffel Tower before it and the Concorde after it—a status symbol of the "Grande Nation".

Historical Design Concepts

One can have different views concerning which historical design concepts could serve as an inspiration for development work on electric cars with intrinsic cultural value; engaging with them, however, seems at least to provide orientation. What is decisive is that this is not done from a nostalgic viewpoint, but from an historical one, i.e. one that takes into account the contemporary context. In this sense, it would make sense to consider automobiles that, on the one hand, meet the criteria cited by Leggewie, and, on the other hand, are conceived as aesthetically integrative products. One such car, for example, was the Renault 4, also from France. This car was conceived from the outset as a French "Volkswagen" and was already a rather small and modestly-powered car when it was introduced in 1961. Thanks to its unusual and at the same time practical operating concept (the "revolver gear change"), it was nevertheless a driving pleasure for its contemporaries. In addition, the intelligent and flexible interior layout offered a high degree of spaciousness. The decisive thing, however, was that the body design of the Renault 4 was—one can say this in retrospect—pictorial in a culturally positive sense, because it touched many contemporaries emotionally, without in the slightest catering to such questionable needs as virility or "top dog" behavior. As a car that for decades people of the most diverse social and class backgrounds were able to relate to, as a consumer good it had a socially unifying rather than a divisive function.[28] The integrative character of the Renault 4 design was by no means a product of chance; on the contrary, it was wholly intentional. In September 1956, Renault boss Pierre Dreyfus had commissioned his developers to create a "folksy and practical car" that would be more "aesthetic" than competing models and "as practical and classless as blue jeans, the garment that can be worn on all occasions, is hard-wearing and can be bought anywhere in the world" (Schrader/Pascal 1999: 15).

Aesthetically Integrative Products

Restrained Internal Forms that Gently Enliven the Bodywork

If one wanted to encapsulate the main aesthetic feature of the Renault 4 in a single phrase, one could say that it is sober and functional without being cold. At first glance, it is "box-like" with rounded corners and edges (a "soft box"), but a closer look reveals a differentiated play of restrained internal forms that gently enliven the bodywork. In addition, there are some very individual design details such as the counter-rotating rhythm of the graphic lines of the side view or the characteristic third side window. The overall composition of the Renault 4 brings together seemingly contradictory elements: the car is visually simple and complex at the same time, one can grasp it immediately and yet repeatedly re-discover it. Symbolically, it represents its owner as a self-confident individual, not as an actor in need of status.[29]

Self-confident Individual, Not an Actor in Need of Status

Lastly, one should mention an Italian car designer who, during the 1970s and 1980s, created a series of cars that could also serve as an inspiration for the design of electric cars today. The designer in question was Giorgetto Giugiaro, who with

---

[28] The Renault 4 was produced from 1961 to 1992. With a total volume of over 8 million units, it is one of the best-selling automobiles ever.

[29] In the words of an R4-driving personnel manager (45): "A certain bizarreness—I don't want to say ugliness—also ensures great individuality and timelessness. R4 design is ultimately good design, because even after 30 years, it doesn't look embarrassing at all." Quoted in an article in *Motor Klassik* magazine, issue 7/2011, p. 64.

his company Italdesign designed such successful models as the VW Golf I (1974), the Fiat Panda (1980) and the Fiat Uno (1981). With these cars the highly innovative practical value has always been emphasized: space for four to five people despite the vehicle's small external dimensions, folding seats and seat back benches, large tailgate, high load capacity, etc. Seemingly less appreciated, on the other hand, are their special aesthetic qualities. As is the case with the R4, they, too, are visually complex in a striking way: ostensibly, they are equipped with a strong expression of sober functionality,[30] but at the same time, thanks to their vivid pictorial qualities, these cars appear more intense and "alive" than most of the products from their competitors at the time—whether this was due to the taut proportions of the bodies, which radiate a discreetly vital physicality, or to the front ends, which look out into the world with self-confident and friendly 'faces'.

Discreetly Vital Physicality

When Giugiaro received these commissions, which can be described as major projects due to the large quantities involved, he had already been in the business for around 20 years. As a designer with an artistic upbringing—his father and grandfather were painters—his talent was discovered and encouraged early on. The first period of his working life was marked by his collaboration in the noble Turin design company, Bertone, which specialized in the development of exclusive cars for top-class brands such as Alfa Romeo, Maserati or Ferrari. At Bertone and during the first years of working independently, Giugiaro worked on a number of sports cars that are now considered design icons—including the BMW M1 mentioned earlier. In this environment, the designer's expressiveness was able to mature into a profound competence that, in the medium term, was not bound to the object "sports cars". In any case, Giugiaro proved this when, in the period after Bertone, he came up with the above-mentioned small cars, attractive across the lines of social class. These are all situated on the same design level as his sports cars, without copying their specific characteristics (Fig. 4.5).[31]

---

[30] The advertising campaign for the Fiat Panda, for example, illustrated this expression in the slogan "The Wonderful Box" ("Die tolle Kiste").

[31] In his study "Criticism of the Car", the influential communication designer Otl Aicher expressly praised Giugaro's achievements: "The Golf was already striking, but the Uno has an overarching aesthetic character. It is even elegant" (cf. Aicher 1984: 46).

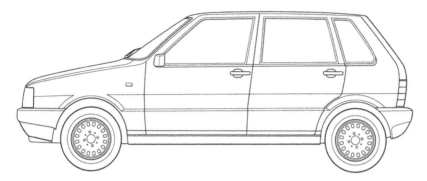

Fig. 4.5 Functional Elegance—The Fiat Uno (Design: Giogietto Giugiaro, 1981) (Source: Keichel)

Vital Functionality

The examination of past success stories able to provide orientation for the present must, of course, be done systematically and on a broader empirical basis. The indications are, however, that sustainable design solutions have an aesthetically integrative character. Their symbolism is able to bring together to a certain extent the different lifestyles of people of different ages, social origins and cultural backgrounds, over a period of time that outlasts aesthetic trends and fashions. "Vital functionality" could perhaps be a term that describes such an integrative symbolism. Thus, it would be conceivable to arrange the various elements of a modern electric car in a visually meaningful way that, while softening the character of the car as an apparatus, does not deny it—abstract motifs of corporeality can certainly play a role, as long as they don't push their way into the foreground and appear to be *appliquéd*. What exactly the formal canon should be like with which such a dialectical motif can be realized is essentially an artistic debate. It has to be conducted anew by the designers under the cultural conditions of their time. If this process is successful, there is the prospect of "completely new" (electric) mobility products, with which users will form a durable and at the same time relaxed bond: owning a vehicle longer and driving it less—that would be a contribution to a new culture of mobility.

The Benchmark Is Still Current Behavior

5

Everyday Experience with the Electric Car from the Users' Point of View

Christine Ahrend and Jessica Stock

Introduction

When we talk about Germany as a location for innovation, we have in mind the research and development activities of industry and science as well as the institutional framework established by political policy. The users of the innovations, on the other hand, are generally not taken into account. Thus, the *Commission of Experts for Research and Innovation* (EFI) of the Federal Government stated in its report that the transformation of the energy system requires "a committed, coordinated effort on the part of all actors" (cf. EFI 2012: 5), but it does not mention the users of innovative, environmentally friendly products. This is surprising in that the success of innovative technologies depends directly on their acceptance and the public's willingness to use them. This applies particularly to consumer-oriented products such as the electric car: if the electric car is to make a significant contribution

---

The original version of this chapter was revised. A correction to this chapter is available at https://doi.org/10.1007/978-3-658-29760-2_8

C. Ahrend
TU Berlin, Berlin, Germany

J. Stock (✉)
DFG Research Training Group "Innovation Society Today: The Reflective Production of the New", Bonn, Germany
e-mail: j.stock@bv-bfw.de

© Springer Fachmedien Wiesbaden GmbH, part of Springer Nature 2021, corrected publication 2021
O. Schwedes, M. Keichel (eds.), *The Electric Car*,
https://doi.org/10.1007/978-3-658-29760-2_5

to reducing CO_2-emissions and reducing the dependence on oil, the demand for electric cars must increase significantly. Electric bicycles have already achieved some success in the course of a temporary "hype" about electric mobility. So far, however, they have only been used as a supplement to the existing choices in means of transport.

Lack of Perspective for Users

Deeply-Rooted Mobility Routines

The acceptance of certain innovations in mobility therefore does not necessarily lead to a reduction in motorised private transport. Not only is the automobile at the heart of one of Germany's most important economic sectors, it is also embedded in a comprehensive transport system and deeply rooted in the mobility routines of many citizens. Changing the engine affects not only the construction and design aspects of the car (cf. the contributions by Wallentowitz, Chap. 6 and Keichel, Chap. 4): the changes in the properties of the car that accompany the change in powertrain also have an effect on its use. Even if the price, range and charging time of an electric car should prove comparable to that of a combustion engine in the medium or long term, from the user's point of view it still has characteristics that differ from those of a car with a combustion engine. In what follows we will use the procurement practices of commercial users to show that the success of the electric car is not determined solely by business and politics, but also depends on individual procurement by the end users—who are, after all, the actual bearers of a potentially new mobility culture! Research into purchasing the electric car and the associated demands on concepts of intermodal transport forms the basis of a transport planning that aims to pave the way for a new mobility culture (Fig. 5.1).

Bearers of a New Mobility Culture!

The empirical data to which we refer here come from the project "ICT-based Integration of Electro-Mobility into the Grid Systems of the Future", funded by the Federal Ministry of Economics and Technology (cf. Ahrend et al. 2011). Commercial users of production vehicles (Smart Fortwo electric drive, Karabag 500 E and Micro Vett Fiorino E) from Berlin and North Rhine-Westphalia were

Fig. 5.1 (Source: Stock)

interviewed for the study. We surveyed a total of 36 business users, including 5 women. The majority of users were between 36 and 55 years old. They had to have used the vehicle for a minimum period of 1 month in order to be selected for the study. At the time of the interview, those we surveyed had used their vehicles for an average of 4 months. A broad spectrum of industries is represented in our sample: telecommunications, logistics, medical technology, automotive experts, Federal ministries, utilities, chemical and pharmaceutical industries, real estate, care services and the hotel industry. The vehicles were used for purposes ranging from field trips and customer service to courier trips and "representative" purposes (participation in press events, events, trade fairs). Data collection took place between August 2010 and January 2011 within the framework of interviews arranged on an individual basis, using a semi-structured guideline.

Commercial Users of Electric Cars

Research into Innovation and the Acquisition of Technology in Everyday Life

Acceptance: A Necessary, but Not Sufficient Condition!

Technical innovations as the motor of social development are a central feature of modern societies (cf. Prahalad and Krishnan 2008). Whether an innovation actually manages to establish itself, however, depends on many factors. Acceptance by potential users is a necessary but by no means sufficient condition for the successful establishment of a new technology. Following the 'Social Construction of Technology' approach (cf. Bijker et al. 2005), the development of technology constitutes rather a protracted social process in which different groups of actors are involved. Technical innovations are not simply constructed on the drawing board and then handed over to the user for application. Rather, there is an interrelationship between the various groups of actors. Technology circulates between engineers, other stakeholders involved in the innovation process and users. In this way, ultimately all groups involved in the process become both technology designers and users. However, the electric car can only become relevant to transport policy if users assign it a relevant role in their everyday practices. The often postulated *acceptance* of the electric car does not yet tell us anything about the willingness of the users to commit themselves to the investment of time, patience and money necessary for the change in their transport behaviour. Rather, the acquisition of technology, understood as an adaptation to the requirements of daily practices, is of particular importance.

The Acquisition of Technology as an Act of Adaptation

By the acquisition of technology we mean a learning process in which users familiarize themselves with new technology and either integrate it into their existing routines or develop new routines. The process of acquiring technology does not begin with the actual use. Even before using it, people get an idea of the technology: drivers of electric cars have expectations regarding the characteristics and possible uses of the electric car, which shape their impression of the vehicle in advance. Finally, the learning process itself requires time and, where appropriate, financial investment on the part of the users, a not insignificant degree of flexibility as well as tolerance of frustration, and a medium- or long-term goal.

Reciprocal Relationship Between the Acquisition of Technology and Processes of Social Diffusion

The acquisition of technology is a concept used by researchers to describe both individual and collective processes.[1] From our point of view, however, collective processes of acquisition are more appropriately described using the term diffusion (or dissemination) of technology. In principle, it is true that the individual acquisition of technology and processes of social diffusion stand in a relationship of reciprocal determination: the diffusion of technology is inconceivable without its acquisition and individual processes of acquisition do not take place independently of the way innovations are perceived collectively. The latter is exemplified by innovation-sceptical social fields, such as bureaucracies, in which it is much more difficult for the individual to successfully integrate a new technology into everyday life (Fig. 5.2).

Fig. 5.2 (Source: Stock)

---

[1] Analogous to the concept of the acquisition of technology, sometimes one also uses the term "socialization of technology" (cf. also Tully 2003: 18ff). This term is intended to abstract more strongly from the individual technology and to place the emphasis on a comprehensive learning process. But we prefer the concept of acquisition in order to emphasize the active, creative use of a technology. Technology by no means prescribes a single possible option for using it. It provides the user with room to maneuver.

The Electric Car as Part of Everyday Patterns of Mobility

Technology must not only be accepted, but also integrated into the "knowledge and behavior repertoire" of users (Rammert 1990: 246). For example, the electric car would have to become part of everyday patterns of mobility, i.e. users would have to include it in their transport choices in order to make it relevant to the market. In order to achieve true significance in transport policy terms, the sensitive phase of integrating this new means of transport into individual routines, i.e. changing patterns of mobility, should be strategically taken into account in the planning processes. This can be done by taking up existing trends that reveal altered, less car-focused traffic behavior and including them in concepts of integrated transport (Fig. 5.3).

Integration of the Electric Car into Everyday Practices

It is only when users with different skills, abilities and motivations 'successfully' acquire the product by integrating it into their everyday practices that it proves its worth and is regarded as operational (cf. also Rammert 2000: 97). But why should users try out new technology at all? Compared to the old technology, new

Fig. 5.3 (Source: Stock)

technology can be more compatible with social developments and trends without the old technology losing its utility value. Users integrate new technology into their everyday lives if it is more compatible with their lifestyle or more compatible with changed circumstances: "Functioning in a technical sense includes functioning in a social sense" (Rammert 2000: 93). This is an aspect that is not being taken seriously enough in current research into electro-mobility. Whether an innovative product such as the electric car is able to overcome the deficits of the conventional automobile can only be examined in detail if the technical innovation is actually available as an initial version and is confronted with different contexts of use. The first groups of users are often pioneers, so-called 'early adopters' (cf. von Hippel 1986; Rogers 2003), especially when the significance and purpose of the technology and its application in networks open to innovation have not yet stabilised, i.e. when the social actors involved in the construction and implementation of the new technology are still negotiating the ways in which it will be used and its areas of application. With regard to the general acquisition of the electric car, however, they belong to the 'early majority', i.e. the group that followed the early pioneers in the 1990s. In contrast to early adopters, who are characterized by a strong intrinsic motivation and overlook the usual "teething troubles" of new technologies, the "early majority" more strongly reflects the mainstream. But while the former are often perceived as "freaks" who operate in a social niche, the latter contribute to preparing the way for further user groups through their active acquisition of technology in everyday situations (cf. Rammert 1990: 248). Various visions of use exist side by side, none of which has yet been able to establish itself in the innovation network of electro-mobility.

Early Adopters and Early Majority

Basis for Transport Policy and Planning Strategies

An analysis of the individual acquisition of the electric car in the context of everyday use provides information on the significance that this technical innovation can have in practice. The basis for transport policy and planning strategies that pursue the goal of a broad and sustainable diffusion of electro-mobility can only be provided by insights into the way in which an innovation in transport is integrated into existing routines of mobility or where the limits of such integration lie, and by determining the consequences for the everyday life of those who use the new technology. In what follows, the process of acquiring a new technology will be demonstrated using the example of the commercial use of electric cars.

Substitution or Innovation: Two Perspectives on the Electric Car and Its Use

The users have a differentiated image of the electric car: as a rule, they are neither unconditionally enthusiastic nor completely opposed to it. The differentiated attitude of the users is reflected in the perspectives of substitution and innovation. Nearly all users can and do adopt both perspectives, which, according to our analyses, in principle reveals the potential of electro-mobility for all users.

Potential Uses of Electro-Mobility

For users who adopt the substitution perspective, the electric car is meant to provide a replacement for the conventional combustion engine, and should be just as good or, if possible, even better. Existing mobility needs are thus to be met in terms of an equivalent:

> Would be more fun [if electric vehicles] had characteristics that came closer to cars with combustion engines, if their capacity for acceleration etc. were better. (Interview 27)

From an innovation perspective, however, the electric car is more than just a substitute for the combustion engine. It becomes an independent innovation with a specific product identity that is experienced aesthetically and sensually by the user:

> Well, yeah, sure, it's a whole different technology. So frankly, I found myself totally reminded of Star Wars. I've always imagined it that way. It was like sliding, [...] those noises, it's insane, the way it feels. (Interview 17)

At present, the substitution perspective dominates among most users. They compare the electric car to conventional vehicles and evaluate its performance mainly in comparison with the latter's characteristics.

Reference to Conventional Vehicles

> "But of course you can see the difference. And of course you will also compare them and say: Here is a series, normal, conventionally driven, here it's electric—what are the advantages and disadvantages of the whole thing then?". (Interview 8)

Thus the electric car is classified as less suitable for everyday use, especially with regard to 'range' and 'charging'. The assessment of suitability for everyday use refers not just to the actual use, but also to the different usage potentials offered by the vehicles.

Low Level of Suitability for Everyday Use

The benchmark is always current behavior. You fill up a car with an 80 litre tank within five minutes, […] then you […] depending on what kind of engine you have in there, you're set for 1,000 km. There you go. The Smart takes eight and a half hours to charge and you have a range of 100 km. It doesn't fit the bill. (Interview 9)

You can reach almost any place with a combustion engine car without any complicated preparations. In addition, it is technically reliable and in this respect does not give users any reason to consider mobility alternatives. The electric car, on the other hand, is not equally easy to use. It has not yet proven its technical reliability from the user's point of view and does not promise to be usable for all purposes and in all situations (cf. Graham-Rowe et al. 2012). The additional planning necessary due to the restrictions prevents a potentially unrestricted use and makes the electric car appear less suitable for everyday purposes. The users also come to the same conclusion when the actual commercial use does not at all exploit the full potential of a combustion engine car and an electric car would be perfectly adequate. The chances of success of the electric car as an innovation are therefore not measured by whether it will generally suffice 'objectively' for the distances travelled on a daily basis. Users measure the practical usefulness of the electric car not just on the basis of real uses, but also against the background of their wishes and conceptions of mobility (cf. Ruppert, Chap. 2). A means of transport must live up to these expectations if it is to be lastingly integrated into everyday routines. This can also be seen in the practical behavior of those users who mainly adopt the substitution perspective. During the pilot project, they already made more frequent use of the combustion engine car, do not intend to buy an electric car at a later date and consequently do not intend to use one permanently. The electric car as an innovation is thus not durably stabilized on the practical level. The acquisition of a conventional automobile clearly influences the evaluation of electric cars: the more users refer to the combustion engine car and its potential usefulness as a comparison, the more critically they judge electric cars. The substitution perspective is reinforced by the Federal Government's publications on motorized individual vehicles.

Users' Wishes and Conceptions of Mobility Are Decisive

The Electric Car as Something Special

But users also notice the innovative aspects of the electric car. In certain situations they adopt the innovation perspective instead of the substitution perspective. The electric car is then no longer perceived solely as a vehicle that deviates undesirably from the combustion engine car, but also as something special (Fig. 5.4). The electric car with its specific characteristics may and should partly differ from the combustion engine car. For example, users think up alternative design concepts.

Redesigning Electric Cars

And today you have the opportunity to redesign cars. [...] But instead we imitate Mr Daimler, who invented a motor and then built it into a horse-drawn carriage and simply omitted the horse. And that's exactly how it is today. The electric motor is installed in a normal car designed for the internal combustion engine, and the battery along

Fig. 5.4 (Source: Stock)

with it. And of course that doesn't make any sense. You can design the car differently today. Thanks to the technology that is available, cars can look quite different. And it's slowly starting to happen that designers are designing an electric car that's only conceivable as an electric car. (Interview 12)

Here positive aspects such as noiselessness and zero emissions are foregrounded, while the technical shortcomings take a back seat. In addition, the electric car is associated with the hope of being able to rethink mobility in the future:

On average, 1.25 people sit in a car. Most cars have five seats and then are driven around with four empty seats, 99 percent of the time. And just because people go on holiday once a year, they need a large car. That's nonsense. On holiday, you can rent a car. You can fly to the holiday location and rent a car locally, or take the train, or whatever. Awareness is slowly gaining ground. (Interview 12)

The Electric Car's Potential for Innovation

The product identity of the electric car has also not established itself enduringly in the innovation perspective. Neither the comparison with existing mobility routines nor the prospect of new forms of mobility were able to convince users to view the electric car as something other than a short-term fun vehicle. Nevertheless, the analysis revealed potentials for innovation that users currently associate with the electric car and which, despite all the perceived disadvantages, lead them to enjoy driving the electric car for business purposes. In the first place, the electric car is innovative because it is new, looks unusual and thus has the potential to attract attention (novelty value).

I find it very pleasant. It's futuristic, and when you drive past people who just notice when it's quiet that you don't make any noise, who then peer at you as you drive past, also because of the conspicuous markings on the car … So it's amusing to drive around with it. I find it very pleasant. (Interview 34)

Novelty Value, Promise of Progress and Rarity Value

In addition, they associate the electric car with the promise that it represents an improvement over the combustion engine, at least in the medium and long term (promise of progress). The electric car is also innovative because of the pleasure it gives users. The increased driving pleasure results from good acceleration, the absence of a clutch, as well as a pleasant, at the same time relaxed use due to the quiet to noiseless operation. The electric car is also an innovation for the environment and quality of life. Users associate the electric car with a means of transport that not only increases their quality of life, but also enables them to be mobile in an environmentally friendly way. Lastly, the electric car is also innovative from the users' point of view because it is still rare (rarity value). Not everyone can buy and drive an electric car just like that. Access is still limited (costs, availability), which means that users themselves have a special role to play and can gain a head start advantage over others in terms of experience and use.

Environmentally-Friendly Mobility

> But if you can say: I've driven one or I'm driving it or it's even parked outside—it's always a feeling of being absolutely special because it's so rare to know and be able to do something special, something to which not everyone has access. (interview)

Users enjoy being pioneers. In this way, the innovation imperative of modern societies is also reflected in the users of electric cars: being a pioneer means being innovative (cf. Hutter et al. 2011).

Integrating the Electric Car into Everyday Life

Expectations Surrounding the Electric Car

The process of acquisition begins prior to actual use. Even before using it, users have developed expectations with regard to the electric car. Existing knowledge about electro-mobility can be seen above all in comparisons with electric cars that have been produced beyond the maturity phase of a series or in small quantities. As an innovation, the electric car has circulated between users and designers since the beginning of auto-mobility (Canzler 1997; Merki 2002; Sachs 1990). For this reason, the commercial users surveyed already have a notion of electric cars, although most of them are quite vague in nature. This can be seen in assessment situations, where users state that they were surprised or disappointed by something.

Enduring Use and Modification of Mobility Routines?

Modification of Mobility Routines

Even though the period of use was limited to the duration of the pilot project, the following questions arose with a view to enduring use: (1) How did the users integrate the electric car into their daily work? (2) Has the electric car found its way into the mobility routines of users and led to changes? (3) Can users imagine a future commercial use and a lasting change in mobility routines?

All the users surveyed already had experience with mobility and had developed mobility routines prior to using the electric car (cf. Ahrend 2002b). For these people, the automobile is of particular importance both professionally and privately. Almost without exception, the users drive their car for work purposes. Other means of transport will only be used if the length of the journey makes the train or aircraft more attractive in terms of travel time and comfort. From the user's point of view, occupational requirements currently rule out any other means of transport apart from the car. For the limited period of the study, users had to integrate the electric car into their everyday working lives and thus break with routines.

Limits of Use

The context of commercial and business use, with specified routes and stipulated working hours, sets limits to the commercial use of the electric car from the outset. Commercial users try out the electric car solely with regard to its ability to functionally replace the combustion engine car. As explained above, the internal combustion engine will maintain its role as first car, since the combustion engine is the better option for ensuring mobility for business purposes. Due to the availability of the combustion engine as an alternative, users were very rarely prompted to question their mobility for business purposes in principle. However, the uncertainties surrounding the use of an electric car, particularly the range restriction, led almost all users to consider which vehicle was appropriate for the planned routes, and this also led to a change in driving style. In order to conserve the battery and potentially increase the available range, a more defensive, energy-saving driving style was discernible among users. Many users also frequently left the heating, air conditioning and radio switched off in order to increase the range even further. However, there was no question of developing new mobility routines at this point. Established mobility routines are only seriously questioned by users if such routines are presented not just as an 'external imposition' (cf. Ahrend 2002a, b), but rather within the framework of (personal) 'crisis situations' and the resulting decisions concerning alternative problem-solving strategies (cf. Schäfer/Bamberg 2008; Wilke 2002: 1). However, the change in driving style is perceived by business and commercial users solely as an 'external imposition'. The new driving style did not lead to a re-thinking of their own mobility practices in general. The users considered the changes that would be necessary to be undesirable. It can be assumed that the users have not definitively changed their driving style and they only do so when using an electric car due to the range restriction, coupled with the long charging time. The following are examples of current changes in mobility behavior that can only be classified as temporary: company users report that when using an electric car they prefer to set out on their journey earlier in order to be able to react adequately to any technical difficulties. Other users optimize the routes they take so that the range of the electric car is sufficient for all their activities. Most users also became accustomed to recharging the electric car at the company charging point whenever they returned to the workplace—unless the remaining capacity was well over 90%.

No Development of New Mobility Routines

Changes in Mobility Behavior

In principle, the choice of means of transport, frequency and duration of use during the test period did not differ significantly from commercial mobility patterns when using a combustion engine car. The electric car with its peculiarities is not acquired with a view to adapting and altering existing mobility routines. Existing routines remain the yardstick by which the use of the electric car has to be measured. One single user constituted an exception to this rule: he adapted his mobility routines to the characteristics of the electric car and intends to continue to do so in the future. He regards intermediate charging, for example, as a basic principle of electro-mobility, which is why he charges his vehicle often, using public charging stations. However, there has been no reduction in the frequency with which he uses the car.

Top-up or En-route Charging as a Basic Principle of Electro-Mobility

It is clear that commercial users prefer to revert to a conventional vehicle in the context of current and prospective electro-mobility rather than rethink and change their mobility routines. The electric car is only integrated into everyday working life if its characteristics make it an ideal choice or if the employer insists on it. From the user's point of view, the search for public charging points, intermediate charging and the advance planning of charging cycles are not compatible with existing routines:

> Yes, but what am I supposed to do while the battery is charging? Should I pick my nose for four hours? […] And mobility, I maintain, is a basic provision, which we can take for granted. […] Today you get in a car as a matter of course and are rather frustrated if you can't drive problem-free all the way to Italy […] because a fan belt or something else is broken. That's simply not the way things are meant to be. And then there is electro-mobility as well. This has nothing to do with the fan belt breaking, of course. But this constant recharging […] of the battery, it's simply rubbish. (Interview 9)

Only in the case of a care service provider could the electric car be used immediately as the main vehicle from the point of view of the users in question—if one disregards the purchase price. In this case, this is due to the fact that the ranges offered by electric cars are more than adequate for the purpose.

In general, users assign an important role to the electric car less for the present than for the future. Because of global challenges, in particular climate change, mobility has to be rethought and reorganised in the future. The combustion engine cannot serve the purpose in the long term, since the technology is dependent on fossil-based raw materials. While current mobility routines are being questioned, changing them is not a consideration. Just as it is a risk for entrepreneurs to invest in innovations when the market is difficult to assess, so is it a risk for these users to invest in the reorganisation of their mobility routines when there are no foreseeable alternatives. There is no certainty of outcome for users who want to go to the trouble of participating in the conversion to a new mobility culture. This is illustrated by the following quote:

> But it's not a reliable car. It's not fully developed. It's not technology that's worth paying this price for monthly, and that robs you of the security of owning a car. Because you never really know if something will go wrong again tomorrow, and you'll have to go back to the repair workshop. (Interview 13)

Mobility Needs to Be Rethought and Reorganised

No Certainty of Outcome in the Transformation of the Mobility Culture

The increasing awareness of climate change as a socially relevant problem does not constitute a personal crisis for current users, to the point that it would cause them to change their own mobility practices. Users do not consider scrutinizing their personal transport behaviour and the specific demands they themselves place on means of transport. The means of transport is supposed to be compatible with the users' way of life, but there is no question of adapting their way of life to the means of transport. For them, the duty to adopt environmentally friendly mobility lies elsewhere—first and foremost with the state, which has to plan the framework to facilitate it. Mobility is supposed to be flexible, uncomplicated and limitlessly available. In this respect, users are not very willing to compromise when it comes to incorporating the electric car into their everyday lives. For them, mobility means the having the potential for mobility at any time—and electric cars cannot (yet) offer this without being incorporated into concepts of integrated transport.

Users Unwilling to Compromise

Conclusions

Finally, a specific area of tension can be identified for commercial users and the current state of electro-mobility, which encompasses the dimensions of the future, the present and potentiality: commercial users are increasingly drawing attention to the overall social significance of the electric car for the global future. They regard their role as pioneers as important, because electric mobility can only be made sustainable through (experimental) application. But even in the present, the electric car has to meet operational requirements and fit into mobility routines. Despite their positive experience of electric mobility, the enthusiasm for their own pioneering role and despite the demonstrated suitability of electric vehicles, commercial users are critical of their current use. Mobility routines are vaguely called into question when it comes to the future, but not at all in the present. The ambivalent experiences with this innovation in the present and the high expectations of electro-mobility as a pioneer of a new mobility culture in the near future will be shaped by individual and commercial requirements. Mobility is supposed to be as limitless and as freely available as possible, now and in the future. Although the electric car and its use are supposed to make it possible to design future mobility in a completely different way, at the same time the users want as few restrictions as possible in their routines and options for moving around (Fig. 5.5).

The electric car is supposed to be innovative—and yet not change anything on the individual level that would result in a change in the usual behaviour. Thus, the stated usability has not yet led to the complete integration of the electric car into everyday business life. On the contrary, users and experts are still discussing the

Fig. 5.5   (Source: Stock)

minimisation of the barriers to use and the opportunities offered by intermodal and multimodal, electro-mobile transport concepts. This, in turn, is neither done jointly nor integrated into research on the undoubted potential of electric vehicles to contribute to mobility concepts designed to minimize traffic.

Conflict Between the Future, the Present and Potentiality

Lack of Communication and Cooperation Between Users and Experts

The study of pioneers made it possible to gain an insight into the opportunities offered by electro-mobility. By differentiating between the substitution and innovation perspectives, it is now possible not only to assess the limitations of current use, but also the potential of future use. We have seen that the majority of commercial users are by no means idealistic users willing to compromise. On the contrary, they are critical of the electric car and its usefulness. The focus is not on the innovation perspective but rather on the substitution perspective and thus on current suitability for everyday use compared to the combustion engine car. Examining the experiences of commercial users of electric vehicles has shown the problems with which they are confronted and how they deal with them (acquisition). The problems we have identified must also be taken into account by future users if the electric car is to become widely accepted (diffusion). In a direct comparison, the electric car is not superior to the combustion engine, if the latter constitutes the desired reference point.

Problems with Acquisition and Diffusion

But the innovation potential also shows that the electric car can have a different product identity because it can be flexibly interpreted. It represents an innovation for the environment and quality of life, offering the specific characteristics of noiselessness and a special driving pleasure, and in light of these characteristics—and conceivably others—the electric car can certainly become competitive (cf. Keichel, Chap. 4).

The future impact of the innovation "electric car" is still not fully foreseeable: there are indications that its use has the potential for far-reaching changes in mobility, yet it could also simply continue along the present path. Transport policy measures aimed at a broad implementation of the electric car must be confronted with the fundamental question concerning the role the electric car should play in transport development: will the electric car merely replace the classic combustion engine (substitution) or should the electric car function as a building block within the framework of new mobility concepts that focus on multi-modality and inter-modality and thereby attempt to reduce motorised individual transport (innovation) (cf. Schwedes, Chap. 3)?[2]

The Electric Car Can Be Interpreted in Different Ways

If the electric car is to serve as the basis for new forms of mobility, this requires users to develop new mobility routines. If, on the other hand, the combustion engine is simply to be replaced by the electric car, users can largely maintain their

---

[2]A debate analogous to the substitution and innovation perspective from the user's point of view touches on the question of whether the technological development of the electric car should go in the direction of "conversion" or "purpose design": i.e., whether an existing combustion engine car should be converted to an electric vehicle, or whether it is necessary to develop a new concept that corresponds to the specific requirements of electric cars (cf. Wallentowitz, Chap. 6).

routines. The latter would mean, however, that existing development trends would continue, such as increasing traffic volumes and spatial and infrastructure planning that is strongly oriented towards cars. Whether the electric car can act as a radical or incremental innovation (innovation or substitution) in the future depends not least on the users themselves. A differentiated analysis of the processes of technology acquisition shows that these must be taken into account if undesirable consequences (e.g. simply replacing the combustion engine) are to be avoided and the opportunities offered by electric cars, as well as their limitations, are to be ascertained as comprehensively as possible. Through their active acquisition of technology, users represent an innovation factor that must be taken seriously. The electric car will therefore not prevail as long as the new technology is not pragmatically integrated into the everyday activities of its users.

Radical or Incremental Innovation?

Serious Innovation Factor: Users

Our analysis has shown that the electric car has not (yet) led to any significant changes in mobility routines among commercial and business users, but that it can be integrated into fleets as a second car or as a supplement. At present, environmental friendliness is not a decisive motive for purchasing one.

However, the fundamental orientation of the users towards innovation indicates that the electric car could become an important element of a comprehensive concept of mobility. This in turn would be an important step towards a new mobility culture.

"Focus Battery"

On the Technical Development of Electric Cars

Henning Wallentowitz

Introductory Remarks

The preoccupation with electric cars seems to be subject to a cycle of about 20 years. In the 1990s, an intensive development phase preceded the current work on tasks relating to electromobility (cf. Schwedes, Chap. 3). The electric car was supposed to go on sale in the USA in 1998. The background was a legislative initiative by the state of California which stipulated that 2% of the vehicles sold there had to be zero-emission vehicles. But even before that, after the first energy crisis at the beginning of the 1970s, very similar issues had been addressed (cf. Dreyer 1973; Weh 1974; Braun et al. 1975). The goal then as now was the development of a city car. During the 1972 Olympic Games in Munich, a BMW 1602 was even built as an electric car to accompany the marathon runners. Electric vehicles have never been developed as intensively as they are at present. This is a result of the politically motivated measures intended to combat climate change and the associated fines imposed on the car industry if the enission limits set by the EU are not met (cf. Wallentowitz et al. 2010).

H. Wallentowitz (✉)
Institute of Automotive Engineering Aachen (IKA), RWTH Aachen University, Aachen, Germany
e-mail: henning.wallentowitz@ika.rwth-aachen.de

© Springer Fachmedien Wiesbaden GmbH, part of Springer Nature 2021
O. Schwedes, M. Keichel (eds.), *The Electric Car*,
https://doi.org/10.1007/978-3-658-29760-2_6

The Electric Car in the 20-Year Cycle

The lectures given back in the 1970s, as well as today, deal with the problems of energy storage and motor development. The vehicles themselves just had to be light. Their technical design did not play a role back then. The distinction between converted internal combustion engine vehicles and so-called "purpose design" electric vehicles did not emerge until the 1990s. This approach was then formulated as a breakthrough for the electric vehicle. In fact, however, this turned out not to be the case for cost reasons, and a real opportunity only arose with the transition to existing vehicle structures, developed for combustion engines.

Considerable progress was made in the so-called "Rügen experiment" carried out between 1992 and 1995, in which 60 electric vehicles were made available to private users. The users drove 1.3 million km. This large-scale trial, organized by the Ministry of Research, brought out the questions that still had to be dealt with by all the participating companies. Electric power steering has become established in the meantime, but it is mainly used for vehicles powered by internal combustion engines. The electric air-conditioning system is still struggling with consumption disadvantages because it reduces the range of battery vehicles. For combustion vehicles, heating systems with PTC elements (PTC resistors that heat up when connected to a voltage source) have recently been developed, the significance of which has now been recognized especially for electric vehicles. In modern electric vehicles today, attempts are beeing made to develop aire conditioning that operates in close proximity to the vehicle occupants in order to keep heat loss to a minimum.

The "Rügen Experiment"

Since the Rügen experiment, electric motors in particular as well as their control systems have developed further. There are also interesting suggestions for transmissions. Little progress has been made with batteries, however, and they still fall far short of the performance requirements for vehicles suitable for everyday use. This has resulted in various application scenarios for electric vehicles, which are currently being controversially discussed.

The progress that has been made and still needs to be made are discussed below in order to identify where action is needed in terms of the practical usefulness and environmental benefits of electrotraction.[1]

---

[1] Michelin Challenge Bibendum 2010 offers a comprehensive and readable treatment of the topic; however it does contain some technical simplifications.

Little Progress in Battery Development

Technical Developments

The technical developments of the electric vehicle cover various areas that are initially independent of each other.

Conversion Design Versus Purpose Design

The discussion on how to develop an electric vehicle is once again in full swing, in the current (third) wave of electromobility. Do you take an existing vehicle whose basic development has already been financed and turn it into an electric vehicle by replacing the fuel tank and the combustion engine with a battery and drivetrain, or do you start from scratch and develop a vehicle that is intended exclusively for an electric drivetrain?

The vehicles in the Rügen experiment were all still conversion vehicles, thus manufactured from the Mercedes 190 series, the BMW 3 Series, the VW Golf and Opel vehicles. For cost reasons, this could not have been achieved otherwise. The BMW 3 Series had high-temperature batteries in the trunk and in the former engine compartment. The electric motor was mounted on the rear axle (see Fig. 6.1).

Crash elements were installed in the front end of the vehicle to significantly increase the occupants' chances of survival in the event of a serious accident. While the high-temperature battery (first sodium-sulphur, then sodium-nickel chloride with about 300 V and 300 °C) was used to drive the motor, the rest of the on-board network remained at 12 V. The vehicle battery was supplied via a voltage transformer from the high-temperature battery. The vane pump of the power steering

Fig. 6.1 BMW electric car as conversion design in the 1990s (Source: Author and 7-forum. com)

system was driven by an electric motor, so it constantly consumed energy from the accumulator.

The High-Temperature Battery in Conversion Design

This BMW was decidedly tail-heavy, the body being heavily laden at the rear. This made special reinforcements necessary. Overall, the participants in the Rügen experiment were very satisfied at the end of the test phase and would have liked to continue driving the vehicles. Even the willingness to purchase one was very high, at 50% (Schlager 2010).

At the same time, BMW-Technik GmbH also developed the first serious Purpose Design electric vehicle (see Fig. 6.2).

During the development of the electric vehicle, it turned out that the car could not be built at a realistic cost. The required volume was lacking, although the production technology was to be designed for small quantities. The solution was to

Fig. 6.2 BMW E1 as purpose-design electric vehicle. (**a**) Electric vehicle BMW E1 (**b**) Unique Mobility synchronous motor 30 kW/30 kg. (**c**) A front end which has incorporated a motorcycle engine as a hybrid (**d**) High-temperature battery under the rear seat (Source: BMW Press and Author)

install a combustion engine. The BMW motorcycle engine was used as front wheel drive for this purpose. At the same time a "hybrid over the road" was created. This car was shown at the 1993 Frankfurt Motor Show. Very similar developments are currently underway. The electric mini as a "converted" and the BMW i3 as a "purpose designed" electric car may be examples of this (cf. Keichel, Chap. 4). With no decision in favour of one or the other, it means the systems remain in competition with each other. Other vehicle manufacturers are also offering solutions. Volkswagen is offering the new Golf 7 both as a conventionally-powered vehicle and as an electric vehicle. At Volkswagen, all model series are to be electrified in future. Other vehicle manufacturers have made similar announcements. These changes are supported by subsidies to vehicle buyers, the costs of which are divided between the government and the vehicle manufactures.

Motors and Their Layout in Electric Vehicles

A further point of discussion is the number and layout of the drive motors in electric vehicles. Before the Rügen experiment, it was always assumed that the combustion engine would be replaced by an electric motor, but this is now no longer the general opinion. In the 1970s it was still being discussed whether the drive should be a DC drive, or perhaps a three-phase drive would be better. A three-phase motor, however, requires a complex control system. The latter only became possible thanks to the further development of semi-conductors which supply the required electricity. These developments have taken place since then and no developer today is considering using the "old" DC shunt motor, even though it is still of interest in terms of cost. In modern electric vehicles, either asynchronous or synchronous machines (which are more efficient) are used (see Fig. 6.3).

Direct Current Versus Three-Phase Current Drive

Because of the electronic control, which is similarly complex in both asynchronous and synchronous machines, these drive motors are more expensive than simple DC machines with commutators. The higher speed capability of asynchronous motors has recently been highlighted as a major advantage. The power, which is the product of torque times speed, is then achieved with a lower torque. The torque of a motor is the "cost driver", i.e. motors with high torque are more expensive than those with low torque.

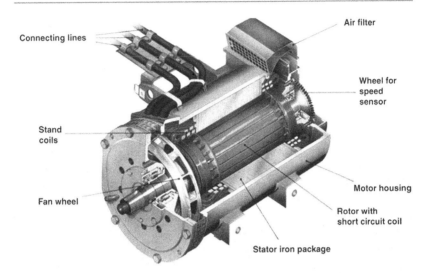

Fig. 6.3 Compilation of motor types. (**a**) Asynchronous motor (Source: Wallentowitz et al. 2010). (**b**) Synchronous motor with permanent magnets. (1) stator. (2) stator backs. (3) copper windings. (4) surface magnets. (5) Rotor core. (6) Rotor shaft in external rotor design. (7) High-voltage plug connection (Source: ZF-Sachs). Anschlussleitungen: Connection Cables; Luftfilter: Air Filter; Rad für Geschwindigkeitssensor: Gearwheel for Speed Sensor; Motorgehäuse: Motor housing; Rotor mit Kurzschlussspule: Rotor with short circuit ring; Ständereisenpaket: Stator core; Lüfterad: Cooling Fan Wheel; Ständerspulen: Three-Phase Windings

"Cost Driver" Torque

Compared to permanently excited synchronous motors, which are only used up to speeds of max. 8000 rpm, asynchronous motors can be operated at over 20,000 rpm, which allows such motors to be made smaller. There are already motors that deliver 90 kW and weigh 45 kg. In this case, however, a multi-stage gearbox is required for its operation. For a motor rotating up to 40,000 rpm, NSK from Japan has developed a gear unit which, for noise reasons, is preceded by a continuously variable gear unit (see Fig. 6.4 and Picture 6.5). This gearbox was presented at the IZB (International Zuliefer-**Börse in** Wolfsburg).

There are currently numerous approaches to the layout of motors in electric vehicles. They can be divided into:

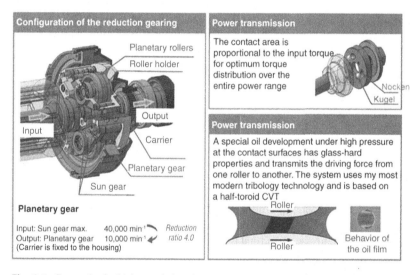

Fig. 6.4   Gear units for high speed electric motors. Konfiguration des Untersetzungsgetriebes: Configuration of the Reduction Gear Unit; Kraftübertragung: Power Transmission; A special oil develops glass-hard properties under high pressure on the contact surfaces and transfers the power from one roller to another. The system uses state-of-the-art tribology technology and is based on a semi-toroidal CVT (continuously variable transmission); The contact surface is proportional to the input torque in order to facilitate optimum torque distribution over the entire power/performance range; Nocken: Cam; Kugel: Ball Bearing; Planetenradgetriebe: Epicyclic gearing/Planetary Gear Unit; Träger: Carrier; Sonnenrad: Sun Gear; Planetenrad: Planetary Gear; Planetenrollen: Planetary Roller; Rollenhalterung: Roll Mounting; Eingang: Input; Ausgang: Output (Source: NSK IZB—Electric Drive System)

Picture 6.5 View of the reduction gear for high speed electric motors Gear for electric motors with up to 40,000 rpm input speed. Advantages: small size, high efficiency and low noise. (Source: Author and NSK IZB—Electric Drive System)

1. central motors
2. near-wheel motors and
3. wheel-hub motors.

Motor Layouts

The central motors act on the front or rear axle in the same way as the combustion engine. At least two separate motors are always used for the near-wheel motors in order to be able to use the advantages of the so-called "torque vectoring". This means that the drive and braking forces can be individually applied to the individual wheels. However, this requires drive shafts that are also subject to high loads. An interesting solution has been proposed by ZF Friedrichshafen AG, where the motors are housed in the trailing arms of a twist beam axle (see Fig. 6.6).

The highest degree of technologization is achieved with wheel hub motors, wherethe electric motor is attached directly to the wheel, thus directly increasing the so-called "unsprung mass". Here, too, there are new solutions, such as the suggestion from NSK in Japan to combine two electric motors via a single gearbox in order to achieve a continuously variable transmission ratio (see Fig. 6.7). Another wheel hub motor was proposed several years ago by the Michelin company. Here

Fig. 6.6 Twist beam axle
with electric motors close to
the wheel (ZF proposal)
(Source: author)

Fig. 6.7 Wheel hub motors
for electric vehicles NSK—
Wheel hub motor for stepless
drive (Source: Author)

the wheel hub motor was additionally combined with the suspension of the vehicle. However, this solution has not yet been implemented in series production, since the motor's utilization of materials is rated as very high.

Overall, the state of the technology shows that considerable progress has been made in the field of electric traction in recent years. While the Unique-Mobility electric motor with permanent magnets in the BMW E1 (see Fig. 6.2), which originated in the USA, still belonged to the exotic category and the Ford Hybrid of 1993 was equipped with a Siemens asynchronous motor (cf. Buschhaus 1994), today's electric vehicles, whether they are "Conversion Design" or "Purpose Design", have at least one such synchronous motor.

State of Development of Batteries

The current view is that only rechargeable batteries can be used as energy storage devices for electric vehicles. The experiments with so-called primary batteries, which were zinc-air batteries that had to be reprocessed and only functioned as exchange systems, were unsuccessful at Deutsche Post 20 years ago. For covering defined distances these batteries were however quite promising. This is probably also the reason why they are now again being considered by developers at the Technical University in Munich. It is not clear, however, that the problems of the past can be regarded as solved today. Battery replacement technology has not really established itself to date (cf. Paluska 2008). In addition, decentralized reprocessing stations must be set up for these so-called "primary batteries", in order to avoid overly long transport distances for the reprocessing.

Primary Versus Secondary Batteries

Rechargeable batteries (so-called secondary batteries) are clearly presented in the Ragone diagram below. The specific power (power density) in W/kg indicates the amount of power that can be obtained from the battery, at least for a short period. This is important for the car's acceleration capability. The specific energy (energy density) in Wh/kg is a measure of the attainable range (see Fig. 6.8).

The shortcomings of these storage devices become fully apparent when liquid fuel data are included in these considerations. For this purpose, the technical data of the energy storage devices are summarized once again in Fig. 6.9. In recent years, the effective range of Li-Ion High Energy batteries has increased considerably. This also depends on the battery design. The design of the cells as push cells, prismatic cells or cylindrical cells plays a role. Significant changes are expected in this area in the future.

Between the energy densities of Li-Ion cells, for example, and gasoline there is a factor of at least 60. Even if the better efficiency for the energy conversion in the electric motor compared to the combustion engine is taken into account, there still remains a factor of 20 in favour of gasoline (or another liquid energy carrier, such as alcohol).

A considerable amount of work remains to be done in the future, if it is actually physically possible to achieve the necessary power densities at all. The ambition is to reach 500 Wh/kg energy density. For the Li-Ion technology, however, this value is not considered achievable today. It follows from these results that for potential

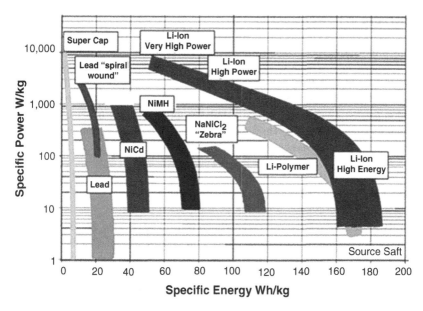

Fig. 6.8 Energy and power density of energy storage devices (secondary batteries) (Source: Saft, France, 2007 and various other sources)

Storage system	energy carriers	energy density		power density	Lifetime/cycle stability	self-discharge
		Wh/kg	Wh/l			
condenser	condenser	4	5	0	+/++	--
secondary cells	NaNiCI	100		-	+/+	-
	Pb-Pb02	20-40	50-100	+	0/0	+
	Ni-Cd	40-60	100-150	+	++/++	-
	Ni-MH	60-90	150-250	+	++/++	--(++)
	Ag-Zn	80-120	150-250	++	-/-	++
	Li-Ion	100-200	150-500	+	+/+	+
fuel	gasoline	12700	8800	++		++
	diesel	11600	9700	++		++

Fig. 6.9 Storage systems at a glance. ++ very good, + good, 0 average, – bad, – – very bad (Source: Praas 2008)

battery alternatives a considerable amount of basic research and application-oriented developments remain necessary. User acceptance can only be significantly increased with optimized ranges (cf. Ahrend and Stock, Chap. 5). The Energy Storage Roadmap of the Fraunhofer Institute for Systems and Innovation Research (2017) also provides a very good overview of energy storage systems.

Increase in Power Density

In addition, however, there are still safety problems with today's battery systems. The "packaging" of battery cells still poses great challenges for system developers. This concerns both the intrinsic safety of the cells, i.e. ensuring that they do not self-ignite, and fire prevention in the event of a crash. Vehicle fires that have occurred in practice point to this necessity. In laboratory tests at the Institute for Motor Vehicles at RWTH Aachen University, the fire hazard resulting from specified damage to the batteries has been confirmed.

The Challenge of Fire Hazards

DEKRA tests have shown in comparisons with gasoline fires that considerably more water is required to extinguish battery fires than to extinguish a fire in a gasoline tank (cf. Dekra 2012). Only with additives To the water are gels formed that make it possible to extinguish a battery fire more easily. We know from work done at the beginning of the 1990s, however, that the fire brigade only has water available. This means that battery fires, especially since they produce toxic gases, remain dangerous. An alternative method for extinguishing fires in electric vehicles that is presently being seriously considered consists in immersing in large, water-filled skips transported on construction vehicles. This is also meant to cool the battery. A further issue that has not yet been resolved industrially is the fire safety of 48 V hybrids. Starting at 20 V electric arcs can be expected in the event of short circuits. The standard fuses can not detect whether it is due to a load or a fault. The problem can be solved by monitoring the 48 V cables by ionisation. To do this, two additional wires have to be installed in the cable, able to detect ionisation (Herges 2015). This problem does not exist to the same extend in high-voltage systems. To avoid or at least reduce the risc of fire, Henkel has been offering technical foams since 2006 (Henkek Tangit), which can be build into the battery boxes.

However, these foams are not yet in use. Prismatic battery containers that slide into each other in the event of an accident (cf. Ginsberg 2011) are now considered state of the art. Overall, it can therefore be said that battery technology represents the greatest obstacle to the establishment of electric vehicles and it has been that way for over 100 years. At present, electric traction is gaining support, especially for two-wheeled vehicles, where the average distances covered are shorter. Electric bicycles can at least be integrated into public transport and, as one means of transport, can support multimodal transport behavior.

100 Years of Battery Technology Without a Breakthrough

The Practical Usefulness of Electrically-Powered Vehicles

Looking at the electric vehicles already in use today, it quickly becomes clear where customers see the practical usefulness. The principal electric vehicles currently in operation are listed in Table 6.1.

This survey makes it clear that the vehicles with a "normal" degree of practical usefulness, i.e. those mentioned under 1., aim for a range of 500 km with one battery charge. The smaller vehicles are only intended for local transport, so thy are more likely to be regarded as second or third vehicles. In recent years, however, their motor performance has also been significantly increased. The ranges strongly depend on the season, the additional options that consume energy, and above all on the handling of the accelerator pedal.

In order to compensate for the disadvantage of limited range, i.e. in order to be independent of battery capacity in terms of range, hybrid vehicles are on offer, which are mainly in the mid-range and executive class. In the future, these vehicles are likely to account for the majority of alternatively-powered cars. Another advantage of hybrids is that the expensive batteries can then be smaller, since the electrical range can be further reduced. This will also reduce the additional costs for hybrid vehicles. The distinction between "Purpose Design" and "Conversion Design" is not necessary for hybrid vehicles anyway, as both drive types have to be accommodated in one vehicle. However, both the Toyota Prius and the Opel Ampera are specially-designed hybrid vehicles, making it easier to accommodate the batteries. However, the return on investment for these vehicles takes longer due to the small quantities produced. This is why the Mercedes and BMW hybrid solutions, which are based on conventional vehicles with internal combustion engines, appear to be more economical. In addition, these vehicles have the full degree of practical usefulness that customers have been accustomed to so far. Whether they really represent economical and ecological solutions for alternative drives is discussed in the following section. Before proceeding, though, let us examing the fourth category in Table 1, fuel cell vehicles. As was the case after the Rügen trials, the industry is still looking for fully-fleged vehicles that can meet the zero-emission requirements. After the Rügen test, various fuel cell vehicles (NECAR 1 from 1994 to NECAR 5 in 2000) were build and extensively tested at Daimler AG. Both hydrogen and methanol were used as fuel. The pressure in the hydrogen tank is specified as 700 bar (10152 psi). However, production of the Mercedes GLC FCEV mentioned in Table 1 will be discontinued in 2020. Other applications for fuel cell are being explored in commercial vehicles and the railway sector (Cuxhaven-Bremerhaven).

Table 6.1 From the vantage point of space on offer, fully-fledged vehicles

1. From the vantage point of space on offer, fully-fledged vehicles	
(a) Nissan Leaf	For 5 people, range 285 km, max speed 144 (km/h)
(b) Nissan Leaf e+	For 5 people, range 385 km, max speed 157 (km/h)
(c) Audi e-tron Quattro 55	For 5 people, range 411 km, max speed 200 (km/h)
(d) Audi e-tron Quattro 50	For 5 people, range 336 km, max speed 190 (km/h)
(e) VW ID-3	For 5 people, range 420 km, max speed 160 (km/h)
(f) VW e-Golf	For 5 people, range 231 km, max speed 150 (km/h)
(g) BMW i3 / i3S	For 4 people, range 359 km, max speed 150 (km/h)
(h) Mercedes EQC 400 4MATIC	For 5 people, range 450 km, max speed 180 (km/h)
(i) Renault ZOE (R 110) u. a.	For 5 people, range 316 km, max speed 135 (km/h)
(j) Tesla Model 3	For 5 people, range 409 km, max speed 225 (km/h)
(k) Peugeot Ion	For 4 people, range 94 km, max speed 130 (km/h)
(l) BYD e5 (China)	For 5 people, range 370 km, max speed 130 (km/h)
(m) Chevrolet Bolt	For 5 people, range 520 km, max speed 145 (km/h)
(n) Opel Ampera e	For 5 people, range 383 km, max speed 145 (km/h)
(o) Hyundai Ionic elec. BEV	For 5 people, range 311 km, max speed 165 (km/h)
(p) Jaguar I-Pace	For 5 people, range 480 km, max speed 200 (km/h)
2. From the vantage point of space on offer: sometimes 2-seaters	
(a) Smart EQ Fortwo	For 2 people, range 160 km, max speed 130 (km/h)
(b) Renault Twizy	For 1–2 people, range 80 km, max speed 80 (km/h)
(c) Renault ZOE Z.E. 50	For 5 people, range 395 km, max speed 135(km/h)
(d) BAIC ArcFox -1	For 2 people, range 200 km, max speed 110 (km/h)
(e) Skoda Citigo - E i V	For 4 people, range 260 km, max speed, 130 (km/h)
(f) Peugeot e-208	For 4 people, range, 340 km, max speed 150 (km/h)
(g) Honda e.	For 4 people, range 200 km, max speed 145 (km/h)
3. Hybrid vehicles	
(a) Toyota Prius PHEV	For 5 people, range 50 km, max speed 135 (km/h)
(b) Mercedes E-Klasse PHEV	For 5 people, range 54 km, max speed 130 (km/h)
(c) Volkswagen Passat GTE	For 5 people, range 66 km, max speed 130 (km/h)
(d) BMW (F30) 330 iPerf.	For 5 people, range 30 km, max speed 120 (km/h)
(e) Volkswagen Tiguan L	For 5 people, range 32 km, max speed 120 (km/h)
(f) Volkswagen Golf GTE	For 5 people, range 60 km, max speed 130 (km/h)
(g) Renault Captur	For 5 people, range 45 km, max speed 135 (km/h)
(h) Hyundai Ioniq HEV	For 5 people, range 3-5 km, max speed 120 (km/h)
4. Fuel cell vehicle	
(a) Honda Clarity FCEV	For 5 people, range 650 km, max speed 165 (km/h)
(b) Toyota Mirai FCEV	For 5 people, range 480 km, max speed 178 (km/h)
(c) Hyundai ix35 FCEV	For 5 people, range 594 km, max speed 160 (km/h)
(d) Hyundai Nexo FCEV	For 5 people, range 666 km, max speed 177 (km/h)
(e) Mercedes GLC FCEV	For 5 people, range 430 km, max speed 160 (km/h)

The supply of hydrogen is also likely to be problematic if the general use of fuel cells in passenger cars is envisaged.

From Electric to Hybrid Vehicles

Comparing the Mass of the Drive Systems

In this section, we assess the influence of the mass of the drive on a vehicle, since it reduces the payload of a vehicle, but it also provides information on an optimum compromise between a battery-powered vehicle and a vehicle powered solely by an internal combustion engine, if hybrid drives are installed. For this assessment, we present the results of a study, the basic findings of which remain valid (cf. Fig. 6.10).

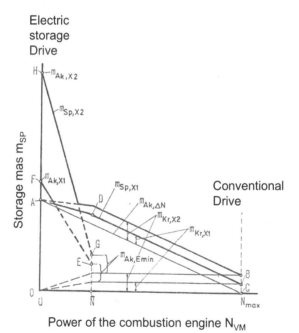

Fig. 6.10 Schematic diagram for determining the storage mass mSP for a hybrid drive system, taking into account the range and the maximum required mileage. Mass of Electrical Storage Device m_{SP}; Performance of Combustion Engine N_{VM}; Battery Drive; Conventional Drive (Source: Dreyer 1973)

The mass of a hybrid drive consists of the individual masses of the storage tank m_{SP} of the generator $m_{gene,}$ of the traction motor $m_{FM,}$ and the combustion engine m_{VM} (including the manual transmission). The graph is designed to show the influence of the storage mass when the criteria "range of the vehicle" and "required driving performance" are weighed against each other. The vertical axis presents the conditions for a purely battery-operated vehicle. On the right side of the diagram the conditions for a purely combustion engine driven vehicle are presented. The battery mass required for a specific vehicle is derived from the required driving performance (e.g. one driving cycle), the vehicle data and the performance characteristics of the battery, as well as the available level of efficiency. **Point A** in the graph is the result of such an estimation. This storage mass also determines the range of the vehicle. For a range X1, the conventionally powered car only needs the amount of fuel $m_{Kr,X1}$. This results in **Point B**, for example. The required fuel mass is derived approximately from the range X, the average driving speed $\bar{v}$, the power $\bar{N}$ and the specific energy of the fuel. The dimensions of the combustion engine determine the maximum achievable driving performance. If the performance of the combustion engine decreases, then the electric drive motor has to deliver the difference in performance and this comes from the electric storage, e.g. the battery.

The mass of the electrical storage device can now $m_{AK, \Delta N}$ be assumed to be inversely proportional to the engine power (assuming constant performance characteristics of the battery). As the power of the combustion engine decreases, there is a linear increase in the battery mass $m_{AK, \Delta N}$ (**line C-A** on the graph). If the installed engine power N_{vm} is less than the average power required for a driving cycle $\bar{N}$ then the originally intended amount of fuel can no longer be completely consumed and the result is the line **D-A**. The purely electric drive requires no fuel at all and is depicted at **point A**.

In addition to the maximum performance currently under discussion, the range is also an important design criterion for the electric vehicle. Using the specific energy of the battery e_{AK} the battery mass $m_{ak,x}$ can be calculated for a given range X. As an example, a range of X1 is depicted by the **point F**.

The Ratio of Battery Mass to Range

In a hybrid drive with a combustion engine that delivers just the average power N required for a driving cycle, an electric storage system—here as a buffer storage— is required in addition to the fuel. **Point E** specifies the required storage mass. How heavy such a storage system has to be depends on the given driving cycle. The difference between the minimum and maximum energy storage indicates the minimum size (and thus the minimum weight) of such a battery with $m_{AK,Emin}$. If the performance of the combustion engine is less than N, the electrical storage system has to provide the difference in energy (**line E-F**).

The Relationship Between Battery Mass and Range

What happens if twice the range is to be achieved is shown by **point H** in the graph. It is assumed that the average values for performance and driving speed remain constant. The energy storage mass of the electric vehicle doubles. For the conventionally or hybrid-powered vehicle, the fuel mass also naturally doubles to $m_{Kr,X2}$ when $N_{VM} = \overline{N}$ is assumed. Under these conditions, the mass $m_{AK,Emin}$ of the battery serving as the storage system is independent of the range. **Point G** on the graph is therefore obtained by shifting the **point E** according to the increased fuel mass ($m_{Kr,X2}$).

In practice, the dimensions of the battery have to be such that the two criteria—range and driving performance—can be achieved jointly. The graph clearly shows that at low power levels of the combustion engine, the range is decisive for the weight of the battery, as soon as the combustion engine can provide at least the average performance required for a driving cycle, the maximum required driving performance is the determining factor.

This also makes it clear why hybrid vehicles will become so interesting in the future. The steep decline in the battery masses (as illustrated by **line H-G** or **line F-E**) show the advantages in terms of weight reduction (and thus also battery costs) when a more powerful combustion engine is used in a hybrid drive.

The Spectrum of Applications for Electric Vehicles

Electric vehicles are often used for so-called commuter services because of their limited range, which, in addition to the long charging time of the battery and the high costs, is stated as a major disadvantage of electric vehicles. So instead of vehicles powered by internal combustion engines, it is better for employees to drive to work in electric vehicles or to use them to cover short distances in the city. Charging stations in shopping centres or at the workplace then facilitate and ensure a continuation of one's journey. The question concerning for whom an electric vehicle is particularly suitable has already been discussed in the past. Detailed descriptions can be found in Wietschel et al. (2012). These assessments show that electric vehicles have to travel greater annual mileages in order to make their purchase worthwhile. Since the financial benefit results from the lower operating costs, but the vehicle itself is significantly more expensive than a vehicle powered by a combustion engine, electric vehicles only make sense if they are actually driven. The above-mentioned commuter applications for city dwellers do not count in this respect. Earlier studies on the high-temperature zebra battery, for example, have shown that, solely considering the amount of energy converted per day, there is a minimum distance that an electric vehicle must cover if its operation is to be economical (cf. Bady 2001). At the time, ten battery electric vehicles were issued to private customers, who used them for their daily mobility. The energy consumed and the daily mileage (averaged from the monthly data) were then plotted on top of each other (see Fig. 6.11). Since the batteries also had to be kept under temperature-controlled conditions during the stand phase (in this instance to maintain them at 300 °C), high energy requirements resulted in low driving performance.

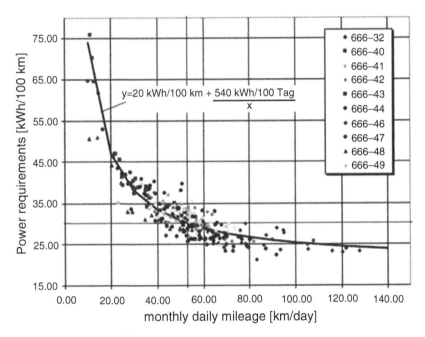

Fig. 6.11 Specific energy consumption ex grid as a function of daily mileage, averaged over a month. Energy requirement (kwh/100 km); Average daily distance travelled over the course of a month (km/day) (Source: Bady 2001)

Field of Application: Commuter Vehicle

Necessity of High Driving Performance

If one compares the electrical energy consumed with the consumption that a vehicle powered by an internal combustion engine would require, one sees that a consumption of 9 l/100 km gasoline or 7.5 l/100 km diesel results in a point of intersection in the range of 65 km/day. At lower mileages, vehicles powered by internal combustion engines are cheaper; at higher mileages per day, electric vehicles are cheaper.

Since this study was carried out, conditions have changed for both internal combustion engines and batteries. The energy requirement for modern lithium-ion batteries, which also need to be heated and cooled, is definitely not as high as for high-temperature batteries. However, the gasoline and diesel consumption of modern vehicles has also fallen significantly. If one assumes 20 kWh/100 km (5 l diesel/100 km or 6 l gasoline/100 km), then the intersection with the consumption curve on the graph would no longer occur. It is thus clear that the high-temperature technology used at the time is in no way promising for the future, at least in terms of cost-effectiveness. Currently there are developers who are returning to this battery (cf. Eimstad 2008). However, there is little likelihood of its ecological success.

If one brings up to date a comparison published in Wietschel et al. (2011) between conventional cars and electric cars with different electricity consumption, one can see in Fig. 6.12 that the so-called "ecological backpack" also demands high overall driving performance from electric cars, even if they are supposed to be ecologically advantageous.

The "Ecological Backpack" of the Electric Car

This comparison shows that in future only electric vehicles powered by renewable electricity will make sense. This is mainly due to the legally enforced improvements in conventional vehicles, which as from 2020 will have to emit less than 95 g CO_2/km.

The so-called ecological backpack is the result of the additional costs involved in the production of the electric vehicle, meaning that, even if it is powered by electricity from renewable sources, the electric vehicle only becomes ecologically more advantageous once it has travelled a distance of more than 75,000 km.

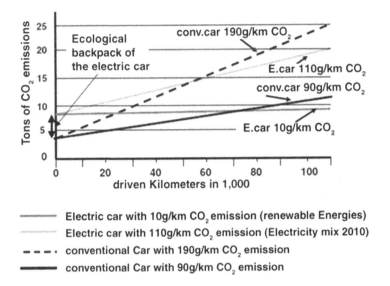

Fig. 6.12 Example of a simplified life cycle assessment of electric vehicles and conventionally powered vehicles. CO_2 emissions in tonnes. kilometres travelled in 1000; electric car with CO_2 emissions of 10g/km (energy from renewable sources); electric car with CO_2 emissions of 110 g/km (electricity from mixed sources); conventional car with CO_2 emissions of 190 g/km; conventional car with CO_2 emissions of 110 g/km (Source: Wietschel et al. 2011 and own calculations)

In summary, it can therefore be stated that electric vehicles need to have high daily mileages in order to use energy economically and that there is a danger that combustion engines optimized in accordance with legal requirements will make life very difficult for electric vehicles. According to certain publications vehicle manufacturers are allegedly considering cutting the production of electric vehicles.[2]

Leisure Time

The relationships between normal mobility and electromobility for general applications, which are quite rightly the object of criticism, contrast with the positive effects of electric vehicles used for leisure purposes. On the list of electric vehicles

---

[2] Under the heading "Opel scraps development of Adam EV", IHS Global Insight reported on 16 November 2012 in its Hybrid EV newsletter on the discontinuation of development activities for the new Opel compact electric car.

available today (see Table 6.1) there are a larger number of smaller vehicles. They are bordered on the one side by quad bikes with a mere 4 kW, which can operated by people 16 years and older with a driver's licence of class S, and on the other side by converted Porsche vehicles with 250 kW. Both have little to do with normal mobility. In between are the vehicles intended for urban use. Since private users do not travel the distances discussed above, it is immediately apparent that other uses have to be found. One of them is likely to be car sharing, which we will come back to later.

Studies have already been done on leisure activities (Dreyer 1973). Even back then it was clear that with a range of 60 km, about 50% of all weekend trips can be done with an electric vehicle. Since ranges have almost doubled in the meantime, but the excursion destinations have remained the same, electric vehicles meet the requirements to an even larger extent.

Great Potential as Vehicles for Leisure Purposes

Use as Commercial Vehicles

Electric vehicles used as commercial vehicles have been around for a long time. This was already the case at the beginning of motorisation, as well as in the 1950s when, for example, the Bundespost in Berlin was still delivering parcels with electric trucks. At the time, lead batteries were the only ones available. With the advent of more flexible small trucks, the parcel service then became simpler and, in principle, cheaper. A new experiment was started at the beginning of the 1990s, when the German Post trialed zinc-air batteries. Since the routes for mail distribution are known in advance, the battery size can be optimized. The power yield of the batteries was promising. The reprocessing of the zinc-air batteries, which are primary batteries, was supposed to be done in the vicinity of the post office buildings. However, the trials carried out did not yield the benefits that had been expected. Zinc-air batteries are only used in a few applications today, e.g. in hearing aids. The development and the construction of the "Street Scooter" e-truck for Deutsche Post, which started in 2010, will be discontinued in 2020. The cost of producing the vehicles in-house is too high.

Other electrically-powered commercial vehicles are trolleybuses, which are still in service in some cities. The main focus here is on zero-emission traffic in cities. In the future, the overhead lines may be replaced by cables set into the roadway, which then transmit the energy inductively to the vehicle. In Korea, intensive research is being conducted into this type of electromobility (cf. Thomson 2010). A relatively good degree of efficiency is achieved by transmitting the current at resonant frequency. The inductive energy transfer process is currently also being developed for contactless charging of electric and hybrid vehicles (plug-in hybrid vehicle) (cf. Wallentowitz et al. 2010). The "emil" city bus in Braunschweig has been inductively charged at several bus stops since 2014.

SUVs (medium-sized commercial vehicles) are fitted out by the company VIA in the USA (Viamotors 2012). Since 2018 there has been a collaboration withh Geely in China. In accordance with the configuration in the Opel Ampera, light commercial vehicles are equipped with an electric drive and a combustion engine. This combination is known as an Extended Range Electric Vehicle (eREV). The vehicle can cover about 40 miles purely electrically (300 kW electric motor). The battery provides 24 kWh and is liquid-cooled. The combustion engine (V6, 4.3 l) supplies 150 kW of electrical power via a flange-mounted generator. This facilitates both direct electric driving and charging of the battery while driving. In addition, external power connections are offered so that tradesmen can use such vehicles as power generators to connect their tools. These solutions are even advertized as emergency power generators for residential buildings. However, as a rule, these vehicles only become economically viable as a result of state subsidies.

Car Sharing

Considerable efforts are currently being made to place electric vehicles in car sharing systems. In this context, car sharing means that customers registered with the system have the locations of available vehicles communicated to them via their mobile phones, which can also be used to open and start the vehicle. The costs are calculated according to the operating time of the vehicles and the kilometres driven. Daimler AG, with its Elektro-Smart, is one of the most committed companies in this field. Other companies are currently following suit, also in cooperation with German Railways (Deutsche Bahn). All major car sharing companies offer electric vehicles in addition to vehicles with combustion engines. DriveNow from BMW and Car2Go from Daimler merged to form ShareNow in February 2019. The number of electric cars should increase significantly throughout Europe by the end of 2019. Flinkster, a subsidiary of Deutsche Bahn, has significantly fewer customers than ShareNow. The proportion of electric cars is low. The vehicles are provided by Flinkster partners in many German cities, often at fixed outlets located near railway stations. WeShare von Volkswagen started in June 2019 with 1500 electric Golfs in Berlin. E-Wald offers its more than 200 electric cars mainly in south-east Bavaria.

That car sharing makes sense follows from the demands made on it, which we have allready outlined: the electric vehicle should be driven as much as possible on a daily basis in order to minimize the losses caused by the vehicle remaining unused, but this requirement is itself limited by the range restriction of the vehicles. Since the individual user only travels short distances, longer daily journeys can only be achieved by having the vehicle used by several car sharing participants. The battery is then recharged during idle times when the vehicles are parked at charging stations. It is not necessary to return the vehicle to a central location, however, since the Smart vehicles could be "collected" by a towing vehicle, then towed simultaneously, one behind the other. Experiments of this kind were already carried out successfully quite a few years ago at the Institute of Automotive Engineering at RWTH Aachen University, since it was already clear at the time that car sharing would be of particular interest, provided there were the freedom of use that has now become a reality (cf. Wallentowitz 2010).

Usefulness of Electric Vehicles in Car Sharing

There are interesting studies on these car sharing activities. For example, car sharing with electric vehicles is conceived as a component of an intermodal mobility offer (cf. Canzler 2011). Accordingly, car manufacturers will turn into mobility service providers in the future. Especially with one-way journeys and when the vehicle does not have to be returned to the rental station, the vehicle utilization is twice as high as with normal car sharing systems (cf. ibid.). This shows how sensible the above mentioned "collecting" of vehicles can be. Daimler AG started a corresponding concept in Ulm in 2008 with the Car2go project.

Utilization Doubled Due to One-Way Car Sharing

Another interesting vehicle suggestion was made by General Motors. This is based on the so-called Segway scooter, where a controller maintains the balance and the rider determines the direction of travel by shifting his weight (cf. Fig. 6.13).

Such a vehicle, well-protected against inclement weather and which can be parked in the smallest possible space (a single axle), would be appropriate for travelling short distances in city traffic, and would thus fit well into a car-sharing system. If the vehicle is equipped with a drawbar, several such vehicles could be collected simultaneously, as described above, requiring only one towing vehicle. However,

Fig. 6.13 Modern proposal for an electric vehicle from GM at the FISITA Congress 2012 in Beijing (Source: Author)

the legal conditions in Germany would have to be changed for this purpose, since no functioning vehicle is permitted to be towed away by another. Perhaps this regulation, which is already a hindrance today, can be altered when it comes to electric mobility, which is surely of great interest to the public.

Conclusion

No Replacement of Existing Vehicle Drives by Electric Motors

The technical development of electric vehicles has reached a high level overall. Powerful motors and transmissions are available as standard, and the units offer solutions that can also be used in vehicles powered by internal combustion engines. However, the performance capabilities of the batteries are still limited, so that there will be no simple replacement of existing vehicle drives by electric motors. However, batteries will also continue to develop in their performance capabilities, as can be seen from the Ragone plot above. Current endavours in Germany by vehicle manufacturers o produce their own batteries could provide a considerable boost to this development. The pressure from the EU, which is threatening to impose fines, is likely to be a further motivation for vehicle manufacturers to get involved. Electromobility will, however, continue to develop in specific fields that do not place great demands on range. It should be borne in mind here that electric vehicles can be used as second vehicles with a high "fun potential" and as car sharing vehicles. Electric vehicles require relatively high daily mileage in order to make economic and ecological sense, because only then will the lower operating costs ensure a return on the investment made in purchasing the vehicles. This also results in special types of mobility, which will only be successful if they allow customers a high degree of flexibility. This means that the focus will not only be on vehicle development as such, but on embedding the vehicles in a user structure (cf. Ahrend and Stock, Chap. 5). According to current and previous findings, this will have an impact above all on communities surrounding the cities.

In Place of an Afterword

Report from an Ongoing Self-experiment

7

Claus Leggewie

Three years ago—I didn't note down the exact date nor did I take a farewell photo—the last automobile I owned was taken away. It was a 1996 silver-blue-grey *Jaguar XJ-6.* The upscale car was already quite old and would have needed an overhaul for a few thousand euros. If my love for automobiles had been as pronounced as half a century earlier, between 1955 and 2005 for example, I would have invested the money without grumbling. The same goes for the sinful purchase price, which is, of course, now exceeded by every 1-year-old Passat. But the looming TÜV (roadworthiness) inspection strengthened the temptation to try out a life (almost) without a car. And to bring the knowledge I had accumulated about the overall harmful effects of gasoline-powered individual transport halfway into line with my actions. As long as the vehicle—which is a pleasure to drive—was parked outside my front door, it's not likely I would have switched to public transport, taxis and the occasional rental car.

C. Leggewie (✉)
Justus Liebig University Gießen, Gießen, Germany
e-mail: claus.leggewie@zmi.uni-giessen.de

Harmonising Knowledge and Actions

My decision was made easier by the fact that my wife owns a car, so I was still able to take the daughter to school in the car and to the usual afternoon events or to the children's birthday party, for example, or to transport heavy loads in the station wagon. But the car wasn't always available for these purposes either, so the bottom line was that, while my life wasn't dramatically disrupted, the change was noticeable.

Gain in Quality of Life

I survived the renunciation and got used to the absence of one of my cars. I'm not missing much, and I'm in the process of acknowledging that there has been a gain in my quality of life. I don't yearn for the Jaguar, I'm hardly likely to buy another car, only occasionally I miss—the CAR.

Initially I would have rated the loss of prestige and reputation higher than the practical difficulties. To think of myself without a car at all seemed to me and my friends hardly possible after such an intensively car-centred life. Since the first Ford 12M (1968), after various sets of wheels in the 1970s (VW Beetle, Renault R16) and a history of sports cars in the 1980s and 1990s (Alfa Romeo Guilia, BMW 525, Jaguar XJ6) it was clear that I didn't use vehicles with internal combustion engines solely and not even primarily as a means of transport, but I identified with them and used them as self-adornment. As I described elsewhere,[1] this went back to the experiences of the post-war period and reconstruction that I share with my whole generation of baby boomers. I liked to use a car for pure driving pleasure, I liked to drive fast (although almost never aggressively) and had myself photographed with my cars (Fig. 7.1). One of my first part-time jobs during my student years was to polish up wheel rims and bumpers at a dealer for luxury sedans and sports cars to promote sales (and take the fancy cars for a spin). I went on pilgrimages to car races, favouring the *Gran Tourismo* sports car category, I hung around in pits and repair workshops for the sights and smells. I could pursue this eulogy—I find it neither strange nor embarrassing. But I lost my close affinity with cars once engine blocks were sealed up so that it was no longer possible to repair anything oneself anyway.

---

[1] *Mut statt Wut. Aufbruch in eine neue Demokratie*, ch. 5 ("Fat Cars"), Hamburg 2011, p. 92 ff.

Fig. 7.1 The author in the VW 1200 Export with folding roof, ca.1961 (Source: Leggewie)

Loss of Relationship as a Result of Abstract Technology

There was still the fun of driving, though. And the fact that it faded away and turned into its opposite has to do with the self-destructiveness of the "unrestricted mobility for free citizens" movement (*freie Fahrt für freie Bürger*)—increasingly annoying were the eternal traffic jams, the never-ending road and bridge repairs, the company representatives roaring up behind you in their Audis and BMWs, wildly flashing their headlights in order to get past. Buying extravagant and good-looking cars for myself also had to do with the fact that I abhorred the streamlined, monotonous models on offer. Anyway: the fun had long since passed me by, for almost all longer journeys I voluntarily switched to the train, which I experienced as consistently stress-free, more reliable and faster. This became the rule when I became a commuter again and had to go on many business trips. Trains are certainly not perfect, but the bottom line is a gain in the quality of travel.

Being a Passenger as a Gain in Quality of Travel

It was much more difficult for me to wait for the bus in bad weather with a heavy bag, to cover short distances on foot or by bicycle, not to spend the money saved on insurance and fuel for taxis and rental cars. Whether I spend more or less since getting rid of the car is something I have deliberately never calculated—saving money was not the main reason for dispensing with the car. Rather, I wanted to try out for myself how difficult it is to practice new forms of mobility. That's why I wasn't interested in hybrid and electric cars, which (as the present book shows) simply replace one drivetrain with another and then continue as before. That's why VW's "blue motion" economy models left me relatively cold, reducing diesel consumption almost to the legendary three-liter level.

New Forms of Mobility

I don't want to be praised or mocked for having (half) done away with my car. It's my business, I don't go around moralizing and absolutely don't see myself as a role model. What was more important to me was that I would draw conclusions from realizing how much I had internalized the automobile as a kind of mental infrastructure and made myself dependent on it. The same applied to the last cigarette a few years earlier, or the introduction to a healthier diet and lifestyle, and giving up other bad habits. As always, this leads to lazy compromises and setbacks. After all, *I'm human.*

Liberation from Dependence: The Car as a Mental Infrastructure

The self-experiment has been anything but perfectly successful, but it's not about perfection, and I don't imagine that I could save the world with the greenhouse gas emissions I've avoided, or even serve as a role model in my milieu. The main aim was to avoid cognitive dissonance. I no longer wanted to be a part of the ridiculous central position of the automobile for the economy, society, politics and culture.

The most difficult task is giving up deeply habitualized individual transport. "Using the car" was and is the *default option.* I still have to think hard about how to get from A to B without the Jaguar (or the replacement car). How will I cope with the possible loss of time and elude the infantile promise made possible by the automobile to have or be able to do everything immediately everywhere? Will I realize what the gains of a "less is more" approach consist of and learn to enjoy the advantages of slowing down and the advantages of the partial renunciation of mobility, which I'm now practicing? Let's see if it works out.

Correction to: The Electric Car

Oliver Schwedes and Marcus Keichel

Correction to:
O. Schwedes, M. Keichel (eds.), *The Electric Car*,
https://doi.org/10.1007/978-3-658-29760-2

The original version of the book was inadvertently published without the following corrections. The chapters have now been corrected.

Corrections:

Page iv: "A subsequent human revision was done primarily in terms of content." has been changed to "The translation was then subjected to a substantial revision for accuracy and style by Gregory Sims."

Page 2, line 9: "Actors from Politics, Research and Industry (Picture 1.1)" has been changed to "Actors from Politics, Research and Industry"

Page 16, line 5: "Dependence on Political and Economic History (Fig. 2.2)" has been changed to "Dependence on Political and Economic History"

The updated online version of this book can be found at
https://doi.org/10.1007/978-3-658-29760-2

Page 18, line 13: "For the former group, the car was partly a machine for sport and pleasure, partly an object that met the requirements of representing status." has been changed to "For the former group, the car was partly a machine for sport and pleasure, partly an object that met the requirements of representing status (Fig. 2.2)."

Page 34, line 8: "Form of Representation (Fig. 2.5)" has been changed to "Form of Representation"

Page 34, line 12: "The glamor of a designer automobile body provided another medium to express the need for distinction of the monied aristocracy and of aristocrats by birth." has been changed to "The glamor of a designer automobile body provided another medium to express the need for distinction of the monied aristocracy and of aristocrats by birth (Fig. 2.5)."

Pages 50, 51 & 52, Table 3.1: The "period (.)" has been included at the end of each cell entries.

Page 104, line 1: "Users' Wishes and Conceptions of Mobility Are Decisive (Fig. 5.4)" has been changed to "Users' Wishes and Conceptions of Mobility Are Decisive"

Page 104, line 6: "The electric car is then no longer perceived solely as a vehicle that deviates undesirably from the combustion engine car, but also as something special." has been changed to "The electric car is then no longer perceived solely as a vehicle that deviates undesirably from the combustion engine car, but also as something special (Fig. 5.4)."

Bibliography

Ahrend, Christine (2002a): Mobilitätsstrategien erforschen. In: Wulf-Holger Arndt (Ed.), Verkehrsplanungsseminar. Beiträge aus Verkehrsplanungstheorie und -praxis. Berlin, S. 63–72.

Ahrend, Christine (2002b): Mobilitätsstrategien zehnjähriger Jungen und Mädchen als Grundlage städtischer Verkehrsplanung. Münster.

Ahrend, Christine/Oliver Schwedes/Jessica Stock/Iris Menke (2011): Ergebnisbericht der Technischen Universität Berlin im Teilprojekt : Analyse Nutzerinnen und Nutzerverhalten und Raumplanung regionale InfrastrukturVerbundprojekt "IKT-basierte Integration der Elektromobilität in die Netzsysteme der Zukunft". Technische Universität Berlin.

Aicher, Otl (1984): Schwierige Verteidigung des Autos gegen seine Anbeter. München.

Allmers, Robert/R. Kaufmann/C. Fritz/E. Kleinrath (Ed.) (1928): Das deutsche Automobilwesen der Gegenwart. Berlin.

Altvater, Elmar/Birgit Mahnkopf (2007): Grenzen der Globalisierung. Ökonomie, Ökologie und Politik in der Weltgesellschaft, 7. Auflage. Münster.

Bady, Ralf (2001): Technisches Einsatzpotential von Elektrofahrzeugen mit Hochtemperaturbatterien im städtischen Alltagsbetrieb. Dissertation RWTH Aachen.

Banister, David (2008): Unsustainable Transport. City transport in the new century. Oxfordshire/ New York.

Barthes, Roland (1964): Mythen des Alltags. Frankfurt M.

Beck, Ulrich (1986): Risikogesellschaft. Auf dem Weg in eine andere Moderne. Frankfurt M.

Becker, Peter (2010): Aufstieg und Krise der deutschen Stromkonzerne. Bochum.

Berger, Roland/Hans-Gerd Servatius (1994): Die Zukunft des Autos hat erst begonnen. Ökologisches Umsteuern als Chance. München/Zürich.

Bierbaum, Otto Julius (1906): Ein Gespräch über das Automobil. In: Mit der Kraft. Berlin, S. 321.

Bijker, Wiebe E./Thomas P. Hughes/Trevor J. Pinch (Ed.) (2005): The Social Construction of Technological Systems. New Directions in the Sociology and History of Technology. Cambridge, Mass.

© Springer Fachmedien Wiesbaden GmbH, part of Springer Nature 2021
O. Schwedes, M. Keichel (eds.), *The Electric Car*,
https://doi.org/10.1007/978-3-658-29760-2

Billisch, Franz Robert/Ernst Fiala/Hans Kronberger (1994): Abenteuer Elektroauto. Frei-
    enbach.
Bode, Peter M./Sylvia Hamberger/Wolfgang Zängl (1986): Alptraum Auto. Eine hundertjäh-
    rige Erfindung und ihre Folgen. München.
Borscheid, Peter (1988): Auto und Massenmobilität, in: Hans Pohl (Ed.): Die Einflüsse der
    Motorisierung auf das Verkehrswesen von 1886 bis 1986 (Tagung 27./28. November
    1986 in Fellbach). Zeitschrift für Unternehmensgeschichte, Beiheft 52. Stuttgart.
Borscheid, Peter (2004): Das Tempo-Virus: Eine Kulturgeschichte der Beschleunigung.
    Frankfurt/New York.
Blume, Jutta/Nika Greger/Wolfgang Pomrehn (2011): Oben Hui, Unten Pfui? Rohstoffe für
    die "grüne" Wirtschaft: Bedarfe – Probleme – Handlungsoptionen für Wirtschaft, Politik
    & Zivilgesellschaft. Berlin.
Braun, Horst/Wilhelm Dreyer/Claus Wolf (1975): Sonderforschungsbereich 97, Fahrzeuge
    und Antriebe, Teilprojekt Stadtkraftfahrzeuge, Bericht 35, Betriebliche Anforderungen
    an Stadtkraftfahrzeuge. TU Braunschweig.
Braungart, Michael/William MCDonough (Ed.) (2011): Die nächste industrielle Revolution.
    Die Cradle to Cradle-Community. Hamburg.
Bundesregierung (2009): Nationaler Entwicklungsplan Elektromobilität der Bundesregie-
    rung. Berlin.
Buschhaus, Wolfram (1994): Entwicklung eines leistungsorientierten Hybridantriebs mit
    vollautomatischer Betriebsstrategie. Dissertation RWTH Aachen.
Burkhardt, Francois (1990): Geschwindigkeit in Gestalt und Fortschritt als Propaganda.
    Streamline und Stromlinie in Amerika und Europa, in: Angela Schönberger (Ed.): Ray-
    mond Loewy. Pionier des amerikanischen Industriedesigns. München.
Canzler, Weert (2011): Vernetzte Mobilität für die Stadt von morgen. Yellow Paper Stadt der
    Zukunft, EMM Europäische Multiplikatoren Medien GmbH. Berlin.
Canzler, Weert (1997): Der Erfolg des Automobils und das Zauberlehrlings-Syndrom, in:
    Meinolf Dierkes (Ed.), Technikgenese. Befunde aus einem Forschungsprogramm. Ber-
    lin, S. 99–129.
Canzler, Weert/Andreas Knie (1994): Das Ende des Automobils. Heidelberg.
Csikszentmihalyi, Mihaly/Eugene Rochberg-Hilton (1989): Der Sinn der Dinge. Das Selbst
    und die Symbole des Wohnbereichs. München-Weinheim.
Dekra (2012): Lithium-Ionen Batterien im Brandversuch. http://www.atzonline.de/Aktuell/
    Nachrichten/1/16907/pr/print.html, Zugriff: 30.09.2012.
Die Bundesregierung (2009): Der Nationale Entwicklungsplan Elektromobilität. http://
    www.elektromobilitaet2008.de, Zugriff: 30.09.2012., Zugriff: 30.09.2012.
DIW – Deutsches Institut für Wirtschaftsforschung (2011): Verkehr in Zahlen 2011/2012.
    Berlin.
Dreyer, Wilhelm (1973): Sonderforschungsbereich 97, Fahrzeuge und Antriebe, Teilprojekt
    Stadtkraftfahrzeug, Bericht 3, Anforderungen an das Antriebssystem eines Stadt-Pkw.
    TU Braunschweig.
Eichberg, Henning (1987): Die Revolution des Automobils. In: Ders.: Die historische Rela-
    tivität der Sachen und Gespenster im Zeughaus. Münster.
EFI – Expertenkommission Forschung und Innovation (2012): Gutachten zu Forschung, In-
    novation und Technologischer Leistungsfähigkeit Deutschlands 2012. Berlin.

Eimstad, Michael (2008): Das elektrische Stadtauto Think City. In: Fährt das Auto der Zukunft elektrisch? Dokumentation der Konferenz vom 28. April 2008 in Berlin.

Fenn, Jackie/Mark Raskino (2008): Mastering the Hype Cycle. How to Choose the Right Innovation at the Right Time. Boston (Massachusetts).

Fünfschilling, Leonhard, Hermann Huber (Ed.): Risse im Lack. Auf den Spuren der Autokultur (Schweizer Werkbund). Zürich 1985.

Geertz, Clifford (1987): Dichte Beschreibung. Beiträge zum Verstehen kultureller Systeme. Frankfurt M.

Ginsberg, Sven (2011): Crashdeformierbares Batteriekonzept für Elektrofahrzeuge, Aachener Karosserietage.

GGEMO – Gemeinsame Geschäftsstelle Elektromobilität der Bundesregierung (2011): Zweiter Bericht der Nationalen Plattform Elektromobilität, Berlin. http://bmwi.de/Dateien/Energieportal/PDF/zweiter-bericht-nationale-plattform-elektromobilitaet,property=pdf,bereich=bmwi,sprache=de,rwb=true.pdf., Zugriff: 30.09.2012.

Gläser, Kai/Christoph Danzer/Rene Kockisch (2012): Radnaher Hochleistungs-Elektroantrieb mit integriertem Planetengetriebe, Fachworkshop Elektromobilität, Zentrales Innovationsprogramm Mittelstand (ZIM), Berlin, 07.04.2012.

Goebbels, Joseph (1939): Rede in: Allmers, Robert /Joseph Goebbels/Adolf Hitler: Kräfte lenken, Kräfte sparen. Drei Reden zur Internationalen Automobil und Motorrad-Ausstellung. Hrsg. vom Reichsverband der Automobilindustrie. Berlin.

Graham-Rowe, Ella/Benjamin Gardner/Charles Abraham/Stephen Skippon/Helga Dittmar/ Rebecca Hutchins/Jenny Stannard (2012): Mainstream consumers driving plugin batteryelectric and plug-in hybrid electric cars: A qualitative analysis of responses and evaluations. In: Transportation Research Part A 46, S. 140–153.

Gropius, Walter (1914): Moderne Probleme der Verkehrsbewegung. In: Jahrbuch des deutschen Werkbundes. Jena.

Haberland, Michael (1900): Das Fahrrad. In: Ders.: Cultur im Alltag, Gesammelte Aufsätze. Wien.

Habermas, Jürgen (1985): Die neue Unübersichtlichkeit. Kleine Politische Schriften V. Frankfurt M.

Haipeter, Thomas (2001): Vom Fordismus zum Postfordismus? Über den Wandel des Produktionssystems bei Volkswagen seit den siebziger Jahren. In: Rudolf Boch (Ed.): Geschichte und Zukunft der deutschen Automobilindustrie. Stuttgart, S. 216– 246.

Herges, Peter (2015): Procedures and Apparatuses to Identify the Electric Arc, European Patent Specification EP 3 161 918 B1, registration 15.6.2015.

Hickethier Knut/Wolf Dieter Lützen/Karin Reis (1974): Das deutsche Auto. Volkswagenwerbung und Volkskultur. Steinbach.

Hippel, Eric von (1986): Lead users: a source of novel product concepts. In: Management Science 32 (7), S.791–805.

Honnef, Klaus (Ed.) (1972): Verkehrskultur. Recklinghausen.

Hoogma, Remco/René Kemp/Johan Schot/Bernhard Truffer (2002): Experimenting for Sustainable Transport. The approach of Strategic Niche Management. London.

Huber, Joseph (1995): Nachhaltige Entwicklung durch Suffizienz, Effizienz und Konsistenz. In: Peter Fritz/Joseph Huber/Hans-Wolfgang Levi (Ed.): Nachhaltigkeit in naturwissenschaftlicher und sozialwissenschaftlicher Perspektive. Stuttgart, S. 31– 46.

Hutter, Michael/Hubert Knoblauch/Werner Rammert/Arnold Windeler (2011): Innovationsgesellschaft heute: Die reflexive Herstellung des Neuen, Discussion Paper TUTS-WP-4-2011. Technische Universität Berlin.

IDW – Institut der Deutschen Wirtschaft (2011): Elektromobilität. Studie zusammen mit der DB Research, http://www.iwkoeln.de/Portals/0/pdf/Elektromobilit%C3%A4t.pdf, Zugriff: 30.09.2012.

IPCC – Intergovernmental Panel on Climate Change (2007): Climate Change 2007: Mitigation of Climate Change, Cambridge University Press. Cambridge/New York.

Jakobs, Eva-Maria/Katrin Lehnen/Martina Ziefle (2008): Alter und Technik. Studie zu Technikkonzepten, Techniknutzung und Technikbewertung älterer Menschen. Aachen.

Katzemich, Nina (2012): Politische Einflussnahme. Die Autolobby in Brüssel. In: Umwelt aktuell 11/2012, S. 4–5.

Kaschuba, Wolfgang (2004): Die Überwindung der Distanz. Zeit und Raum in der europäischen Moderne. Frankfurt M.

Kemfert, Claudia (2013): Kampf um Strom. Mythen, Macht und Monopole. Hamburg.

Knie, Andreas (1997): Die Interpretation des Autos als Rennreise limousine: Genese, Bedeutungsprägung, Fixierungen und verkehrspolitische Konsequenzen. In: Hans Liudger Dienel/Helmuth Trischler (Ed.): Geschichte der Zukunft des Verkehrs. Verkehrskonzepte von der Frühen Neuzeit bis zum 21. Jahrhundert. Frankfurt M./New York, S. 243–259.

Krämer-Badoni, Klaus/Herbert Grymer/Marianne Rodenstein (1971): Zur sozio-ökonomischen Bedeutung des Automobils. Frankfurt M.

Leggewie, Claus (2011): Mut statt Wut. Aufbruch in eine neue Demokratie. Hamburg.

Lessing, Hans-Erhard (2003): Automobilität. Karl Drais und die unglaublichen Anfänge. Leipzig.

Lichtenstein, Claude/Franz Engler (1992): Stromlinienform. Katalogbuch zur gleichnamigen Ausstellung, Museum für Gestaltung, Zürich: 23. Mai bis 2. August 1992.

Linder, Wolf/Ulrich Maurer/Hubert Resch (1975): Erzwungene Mobilität. Alternativen zur Raumordnung, Stadtentwicklung und Verkehrspolitik. Köln/Frankfurt M.

Linzbach, Antonia/Joris Luyt/René Krikke (2009): Electric Cars. An Assessment of the Stabilization of the Electric Car in Europe Using SCOT Theory. Enschede.

Lützen, Wolf Dieter (1986): Radfahren, Motorsport, Autobesitz. Motorisierung zwischen Gebrauchswerten und Statuserwerb. In: Wolfgang Ruppert (Ed.): Die Arbeiter. München.

Maak, Nicklas (2012): Die kalte und die heiße Stadt. In: TU München und Bayrische Akademie der Schönen Künste (Ed.): Die Tradition von morgen. Architektur in München seit 1980. München.

Marx, Karl/Friedrich Engels (1972): Werke, Band 4. Berlin, S. 459– 493.

Merki, Christoph M. (2002): Der holprige Siegeszug des Automobils. 1895–1930. Zur Motorisierung des Straßenverkehrs in Frankreich, Deutschland und der Schweiz. Wien.

Meyers Großes Konversationslexikon, 6. Print, 40. Vol. Leipzig and Wien.

Michelin Challenge Bibendum (2010): Mobilität morgen. Der nachhaltige Straßenverkehr der Zukunft. Paris.

Möser, Kurt (2002): Geschichte des Autos. Frankfurt/New York.

Mom, Gijs (2011): Avantgarde – Elektroautos um 1900. Vortrag im Rahmen der Reihe Auto. Mobil.Geschichte der Universität Stuttgart, am 15. Mai 2011, in Stuttgart. http://www.stuttgart.de/item/show/432537, Zugriff: 10.01.2013.

Mom, Gijs (2004): The Electric Vehicle. Technology and Expectations in the Automobile Age. Baltimore/London.

Mommsen, Hans (1996): Das Volkswagenwerk und seine Arbeiter im Dritten Reich. Berlin.

Öko-Institut (2011): Autos unter Strom. Ergebnisbroschüre erstellt im Rahmen des Projektes OPTUM "Umweltentlastungspotentiale von Elektrofahrzeugen Integrierte – Betrachtung von Fahrzeugnutzung und Energiewirtschaft". Berlin.

Paluska, Joe (2008): Das Projekt Better Place. In: Fährt das Auto der Zukunft elektrisch? Dokumentation der Konferenz vom 28. April 2008 in Berlin.

Petersen, Rudolf (2011): Mobilität für morgen. In: Oliver Schwedes (Ed.): Verkehrspolitik. Eine interdisziplinäre Einführung. Wiesbaden, S. 411–430.

Petsch, Joachim (1982): Geschichte des Auto-Design. Köln.

Peukert, Helge (2011): Die große Finanzmarkt- und Staatsschuldenkrise. Eine kritisch-heterodoxe Untersuchung. 2. Auflage. Marburg.

Polanyi, Karl (1995/1944): The Great Transformation. Politische und ökonomische Ursprünge von Gesellschaften und Wirtschaftssystemen. Frankfurt M.

Polster, Bernd (1982): Tankstellen. Die Benzingeschichte. Berlin.

Praas, Hans-Walter. (2008): Technologische Grundlagen moderner Batteriesysteme, Basiswissen Batterie, Vortrag TAE Workshop. Esslingen.

Prahalad, C.K./M.S. Krishnan (2008): The new age of innovation. New York.

Princen, Thomas (2005): The Logic of Sufficiency. Cambridge(Massachusetts)/London.

Radkau, Joachim (2011): Die Ära der Ökologie. Eine Weltgeschichte. München.

Rammert, Werner (1990): Telefon und Kommunikationskultur. Akzeptanz und Diffusion einer Technik im Vier-Länder-Vergleich. In: Kölner Zeitschrift für Soziologie und Sozialpsychologie 42, S. 20–40.

Rammert, Werner (2000): Technik aus soziologischer Perspektive 2. Kultur – Innovation – Virtualität. Wiesbaden.

Rogers, Everett M. (2003): Diffusion of Innovations. New York.

Ruppert, Wolfgang (1993): Das Auto. Herrschaft über Raum und Zeit. In: Wolfgang Ruppert (Ed.): Fahrrad, Auto, Fernsehschrank: Zur Kulturgeschichte der Alltagsdinge. Frankfurt M.

Ruppert, Wolfgang (1998): Der moderne Künstler. Zur Sozial- und Kulturgeschichte der kreativen Individualität in der kulturellen Moderne im 19. und frühen 20. Jahrhundert. Frankfurt M.

Sachs, Wolfgang (1990): Die Liebe zum Automobil. Ein Rückblick in die Geschichte unserer Wünsche. Reinbek bei Hamburg.

Schäfer, Martina/Sebastian Bamberg (2008): Breaking Habits: Linking Sustainable Consumption Campaigns to Sensitive Life Events. Proceedings: Sustainable Consumption and Production: Framework for Action. Conference of the Sustainable Consumption Research Exchange (SCORE!) Network, supported by the EU's 6th Framework Programme. Brüssel.

Scheer, Hermann (2010): Der Energethische Imperativ. 100 % jetzt: Wie der vollständige Wechsel zu erneuerbaren Energien zu realisieren ist. München.

Schier, Michael (2010): Überblick über Elektroantriebe, Workshop FVEE, 20.01.2010, Ulm, Institut für Fahrzeugkonzepte des DLR in Stuttgart.

Schindler, Jörg/Martin Held/Gerd Würdemann (2009): Postfossile Mobilität. Wegweiser für die Zeit nach dem Peak Oil. Bad Homburg.

Schivelbusch, Wolfgang (1977): Geschichte der Eisenbahnreise. Zur Industrialisierung von Raum und Zeit im 19. Jahrhundert. München/Wien.

Schlager, Katja (2010): Kundenerwartungen an die Elektromobilität, Anwenderforum Mobi-
    liTec, 20. April 2010, Hannover Messe, Institut für Transportation Design (ITD) Braun-
    schweig.
Schmucki, Barbara (2001): Der Traum vom Verkehrsfluss. Städtische Verkehrsplanungseit
    1945 im deutsch-deutschen Vergleich. Frankfurt/New York.
Schrader, Haltwart/Dominique Pascal (1999): Renault – Vom R4 zum Kangoo. Stuttgart.
Schumann, Eric (1981): Vom Dampfwagen zum Auto. Motorisierung des Verkehrs. Rein-
    beck.
Schumpeter, Joseph A. (1950/1942): Kapitalismus, Sozialismus und Demokratie. Tübingen.
Schwedes, Oliver/Martin Gegner (2013): Der Verkehr des Leviathan – Zur Genese des städ-
    tischen Verkehrs im Rahmen der Daseinsvorsorge. In: Oliver Schwedes (Ed.): Öffent-
    liche Mobilität. Perspektiven für eine nachhaltige Verkehrsentwicklung. Wiesbaden (in
    Vorbereitung).
Schwedes, Oliver/Stephan Rammler (2012): Mobile Cities. Dynamiken weltweiter Stadt-
    und Verkehrsentwicklung. Berlin.
Schwedes, Oliver/Chistine Ahrend/Stefanie Kettner/Benjamin Tiedtke (2011a): Elektromo-
    bilität – Hoffnungsträger oder Luftschloss. Eine akteurszentrierte Diskursanalyse über
    die Elektromobilität 1990 bis 2010. http://www.verkehrsplanung.tu-berlin.de http://
    www.verkehrsplanung.tu-berlin.de
Schwedes, Oliver/Christine Ahrend/Stefanie Kettner/Benjamin Tiedtke (2011b): Die Geneh-
    migung von Ladeinfrastruktur für Elektroverkehr im öffentlichen Raum. Policy-Analyse.
    http://www.verkehrsplanung.tu-berlin.de
Sennet, Richard (1991): Civitas. Die Großstadt und die Kultur des Unterschieds. Frankfurt M.
Simmel, Georg (1903): Die Großstädte und das Geistesleben. In: Die Großstadt. Vorträge
    und Aufsätze zur Städteausstellung. Dresden.
Steffen, Katharina (1990): Übergangsrituale einer automobilen Gesellschaft. Frankfurt M.
Stengel, Oliver (2011): Suffizienz. Die Konsumgesellschaft in der ökologischen Krise. Mün-
    chen.
Stommer, Rainer (Ed.) (1984): Reichsautobahn. Pyramiden des Dritten Reiches. Analysen
    zur Ästhetik eines unbewältigten Mythos. Marburg.
Thomson, Anthony (2010): Commercial development and deployment of wireless inductive
    power transfer. International Symposium of Futuristic Electric Vehicle, Seoul, Korea, 5.
    April 2010.
Tully, Claus J. (2003): Mensch – Maschine – Megabyte. Technik in der Alltagskultur: Eine
    sozialwissenschaftliche Hinführung. Lehrtexte Soziologie. Opladen.
Veblen, Thorstein (1986): Theorie der feinen Leute. Eine ökonomische Untersuchung der
    Institutionen. Frankfurt M.
Vester, Frederic (1990): Ausfahrt Zukunft. Strategien für den Verkehr von morgen. Eine Sys-
    temuntersuchung. München.
Viamotors (2012): www.viamotors.com/powertrain/. Zugriff: 20.12.2012.
Voy, Carsten (1996): Erprobung von Elektrofahrzeugen der neuesten Generation auf der Insel
    Rügen und Energieversorgung für Elektrofahrzeuge durch Solarenergie und Stromtank-
    stellen. Abschlussbericht, Förderkennzeichen TV 9225 und 0329376A. Braunschweig.
Wallentowitz, Henning (2010): Vom Statussymbol zum Gebrauchsgegenstand. In: Antonio
    Schnieder/Tom Sommerlatte (Hrsg): Die Zukunft der deutschen Wirtschaft. Visionen für
    2030. Erlangen.

Wallentowitz, Henning/Arndt Freialdenhoven/Ingo Olschewski (2010): Strategien zur Elektrifizierung des Antriebstranges, Technologien, Märkte, Implikationen. Wiesbaden.

Warnstorf Berdelsmann (2012): Trendstudie Elektromobilität 2012. Hann. Münden/Bremerhafen.

Weh, Herbert (1974): Problematik der Energiespeicher und des elektrischen Antriebs. TU Braunschweig.

Weizsäcker, Ernst Ulrich von/Karlson Hargroves/Michael Smith (2010): Faktor Fünf: Die Formel für nachhaltiges Wachstum. München.

Weyer, Johannes (1997): Vernetzte Innovationen – innovative Netzwerke. Airbus, Personal Computer, Transrapid. In: Werner Rammert/Gotthard Bechmann (Ed.): Technik und Gesellschaft. Jahrbuch 9, Innovation: Prozesse, Produkte, Politik. Frankfurt/M., S. 125–152.

WI – Worldwatch Institute (2012): Vital Signs 2012. Washington.

Wietschel, Martin/Elisabeth Dütschke/Simon Funke/Anja Peters/Patrick Plötz/Uta Schneider/Anette Roser/Joachim Globisch (2012): Kaufpotentiale für Elektrofahrzeuge bei sogenannten "Early Adoptern". Studie des Fraunhofer Institutes für System- und Innovationsforschung im Auftrag des Bundesministeriums für Wirtschaft und Technologie (BMWi). Karlsruhe.

Wietschel, Martin/David Dallinger/Claus Doll/Till Gnann/Michael Held/Fabian Kley/Christian Lerch/Frank Marscheider-Weidemann/Katharina Mattes/Anja Peters/Patrick Plötz/Marcus Schröter (2011): Gesellschaftspolitische Fragestellungen der Elektromobilität, Studie des Fraunhofer Institutes für System und Innovationsforschung im Rahmen der Fraunhofer Systemforschung Elektromobilität. Gefördert vom Bundesministerium für Bildung und Forschung. Karlsruhe.

Wilke, Georg (2002): Neue Mobilitätsdienstleistungen und Alltagspraxis, Wuppertal Papers 127. Wuppertal.

Zeller Reiner (Ed.) (1986): Das Automobil in der Kunst 1886 bis 1986. München.

ZF-Sachs: http://www.hybrid-autos.info/Technik/E-Maschinen/aussenler-synchronmaschine-mit-oberflenmagneten.html, Zugriff: 20.12.2012.

Newspaper and Magazine Articles

ADAC Motorwelt, Heft 9, September 2012, Die Wut am Steuer

Auto Motor und Sport, Heft 13, 2012, Des einen Freud, des anderen Light

Der Spiegel, 26.04.1999, Energie der Moderne

Der Spiegel, 08.07.1991, Kommt das Öko-Auto?

Design Report, Heft 3, 2012, Für morgen denken

Frankfurter Allgemeine Zeitung, 12. Januar 2012, Das Leben, vom Tode her gedacht

Frankfurter Rundschau, 06.05.1995, Elektrische Leihmobile gegen die Parkraumnot

Frankfurter Rundschau, 11.05.1996, Zink-Luft-System im Test

Motor Klassik, Heft 7, 2011, Marie und Jeannette

Süddeutsche Zeitung, 15. Februar 2010, "Man bekommt nicht 700 Millionen Dollar für ein Lächeln"

Süddeutsche Zeitung, 13. Oktober 2012, Batterie leer

Süddeutsche Zeitung, 12./13. Januar 2013, Strom aufwärts

Printed in the United States
By Bookmasters